© 2024 by FAISAL JAMIL. All rights reserved.

Title: "The Role of Renewable Energy in Combating Climate Change"

This book, along with its contents encompassing text, illustrations, images, diagrams, and other creative elements, is the exclusive property of FAISAL JAMIL and is safeguarded by copyright law.

FAISAL JAMIL asserts full ownership and retains all rights to this book. No part of this publication may be reproduced, distributed, or transmitted in any form or by any means, such as photocopying, recording, or electronic methods, without prior written consent from the copyright holder. Brief quotations in critical reviews and certain noncommercial uses permitted by copyright law are exceptions.

This copyright notice applies to all editions, formats, and translations of the book, whether in print, digital, or any other medium or technology existing now or developed in the future. Unauthorized use or infringement may result in legal action and pursuit of remedies under applicable copyright laws.

While efforts have been made to ensure accuracy and reliability, FAISAL JAMIL does not guarantee the completeness or suitability of the information. Readers are responsible for evaluating and using the content judiciously.

FAISAL JAMIL reserves the right to make changes, updates, or corrections to the book without prior notice. Inclusion of third-party materials or references does not imply endorsement or affiliation unless used under fair use principles or with proper permissions and attributions.

For permissions, inquiries, or requests regarding the book's use, please contact FAISAL JAMIL through official channels listed on their Amazon author page or provided email address.

This comprehensive copyright notice serves to protect FAISAL JAMIL'S intellectual property rights, maintain content control, and inform users about associated restrictions and permissions.

Warm regards,

FAISAL JAMIL

For Your Feedback and Reviews!

http://www.amazon.com/author/faisal.jamil

Email: faisaljamilauthor@gmail.com

About the author

Certainly! Faisal Jamil is a multifaceted individual with a diverse set of skills and experiences. With a strong foundation in computer knowledge since childhood, he has developed a deep understanding of technology that informs his work as a content writer. Faisal also possesses digital skills, which further enhance his abilities in various digital platforms and technologies.

Beyond his professional endeavors, Faisal Jamil has also excelled in the martial arts, particularly Shotokan Karate, where he achieved the prestigious rank of first Dan black belt. This achievement speaks to his dedication, discipline, and commitment to personal growth and mastery.

In his professional life, Faisal Jamil has carved out a successful career in sales management within the Fast Moving Consumer Goods (FMCG) sector. His roles in various FMCG companies have honed his skills in strategic planning, team leadership, and business development. Faisal's ability to drive sales and achieve targets has been instrumental in his career progression, showcasing his talent for identifying opportunities and delivering results.

Faisal Jamil is also deeply interested in business investment strategies, planning, and execution. His understanding of these areas has been key to his success in the business world, allowing him to make informed decisions and implement effective strategies. His ability to navigate the complexities of investment planning and execution has set him apart as a strategic thinker and a valuable asset in any business endeavor.

Overall, Faisal Jamil is a dynamic individual who combines his passion for technology, martial arts, sales management, digital skills, and business investment strategies to achieve success in diverse fields. His journey is a testament to his versatility, resilience, and continuous pursuit of excellence.

Yours Sincerely

FAISAL JAMIL

For Your Feedback and Reviews!

https://www.amazon.com/author/faisal.jamil

Email: faisaljamilauthor@gmail.com

THE ROLE OF
RENEWABLE ENERGY
IN COMBATING
CLIMATE CHANGE

Table of Content

Preface --8

Introduction ---11

Chapter 1: Introduction to Renewable Energy --------------16

Chapter 2: Historical Perspective -------------------------------22

Chapter 3: The Science of Climate Change -------------------28

Chapter 4: Solar Power: Harnessing the Sun ----------------34

Chapter 5: Wind Energy: Power from the Air ---------------42

Chapter 6: Hydro Power: Energy from Water ---------------50

Chapter 7: Geothermal Energy: Heat from the Earth -----57

Chapter 8: Biomass: Organic Energy --------------------------65

Chapter 9: Energy Storage and Grid Integration -----------72

Chapter 10: Policy and Legislation -----------------------------81

Chapter 11: Economic Impacts ----------------------------------91

Chapter 12: Environmental and Social Benefits ------------99

Chapter 13: Case Studies: Successful Renewable Energy Projects ---107

Chapter 14: Technological Innovations ---------------------114

Chapter 15: Barriers to Adoption -----------------------------125

Chapter 16: The Role of Corporations and Industry ------131

Chapter 17: Community and Grassroots Movements ---138

Chapter 18: The Future of Renewable Energy -------------146

Chapter 19: Renewable Energy and

Developing Nations --152

Chapter 20: Conclusion and Call to Action -----------------159

Closing of the book --167

Thanks you for reading ---------------------------------------170

Preface

As the world grapples with the pressing challenges of climate change, the transition to renewable energy has emerged as a beacon of hope. This book, "The Role of Renewable Energy in Combating Climate Change," aims to shed light on the pivotal role that renewable energy plays in addressing one of the most critical issues of our time. It is a comprehensive exploration of how harnessing the power of the sun, wind, water, and other natural resources can pave the way for a sustainable and resilient future.

The idea for this book was born out of a profound sense of urgency. The scientific consensus is clear: human activities, particularly the burning of fossil fuels, are driving climate change at an unprecedented rate. The impacts are already being felt worldwide, from rising sea levels and extreme weather events to the loss of biodiversity and threats to human health. The need for action is immediate and unequivocal.

Renewable energy offers a viable and scalable solution to mitigate these impacts. Unlike fossil fuels, renewable energy sources are abundant, inexhaustible, and environmentally friendly. They have the potential to transform our energy systems, reduce greenhouse gas emissions, and foster economic growth. This book is a call to understand, embrace, and accelerate this transformation.

In the following chapters, we delve into the various facets of renewable energy. We begin with an introduction to the fundamental concepts, providing a foundation for understanding the different types of renewable energy sources and their applications. From there, we trace the historical evolution of energy consumption and the gradual shift towards cleaner alternatives. The science of climate change is explained in detail, highlighting why renewable energy is crucial in mitigating its effects.

Each chapter is dedicated to a specific type of renewable energy, exploring its technologies, benefits, challenges, and potential. We cover solar, wind, hydro, geothermal, and biomass energy, providing insights into their current state and future prospects. The book also addresses critical issues such as energy storage, grid integration, policy frameworks, economic impacts, and technological innovations.

One of the unique features of this book is the inclusion of case studies from around the world. These real-world examples illustrate how renewable energy projects are being successfully implemented, showcasing the tangible benefits they bring to communities and the environment. Additionally, we examine the barriers to adoption and the innovative solutions being developed to overcome them.

The role of various stakeholders—governments, corporations, industries, and grassroots movements—is explored in depth, emphasizing the collaborative efforts required to drive the renewable energy revolution. The book also looks at the specific challenges and opportunities for renewable energy in developing nations, highlighting its potential to improve lives and foster sustainable development.

As we conclude, we offer a forward-looking perspective on the future of renewable energy. The final chapter summarizes the key points discussed and provides a call to action for individuals, businesses, and governments to take part in this critical endeavor.

This book is intended for a diverse audience, including students, researchers, policymakers, business leaders, and anyone interested in the intersection of energy and climate change. It aims to inform, inspire, and empower readers to contribute to the global transition to renewable energy.

The journey towards a sustainable energy future is not without its challenges, but it is a journey we must undertake. The stakes are high, but so are the rewards. Together, we can harness the power of renewable energy to build a brighter, cleaner, and more resilient world for future generations.

Thank you for joining us on this important journey.

Sincerely,

FAISAL JAMIL

Introduction

Climate change stands as one of the most formidable challenges of our era, posing significant threats to ecosystems, economies, and communities worldwide. The urgency to address this global crisis has never been greater, and central to this endeavor is the transformation of our energy systems. This book, "The Role of Renewable Energy in Combating Climate Change," delves into how renewable energy sources can lead the way in mitigating climate change and fostering a sustainable future.

The Climate Crisis

The evidence of climate change is all around us: rising global temperatures, shrinking ice caps, increasingly severe weather events, and shifts in wildlife populations and habitats. The primary driver of these changes is the accumulation of greenhouse gases (GHGs) in the atmosphere, primarily from the burning of fossil fuels for energy. Carbon dioxide (CO_2), methane (CH_4), and other GHGs trap heat, leading to a warming planet and a host of associated impacts.

The consequences of inaction are dire. Prolonged droughts, intense hurricanes, and sea-level rise threaten food and water security, displace communities, and disrupt economies. Biodiversity loss undermines ecosystems' ability to provide essential services, and human health is compromised by increased air pollution and heat stress. To avert the worst outcomes, it is imperative to drastically reduce GHG emissions and transition to a sustainable energy system.

The Promise of Renewable Energy

Renewable energy offers a path forward. Unlike fossil fuels, renewable energy sources—such as solar, wind, hydro, geothermal, and biomass—are abundant, clean, and sustainable. They provide a means to generate electricity, heat, and transportation fuels without emitting GHGs. By tapping into these natural resources, we can decarbonize our energy systems and build a resilient, low-carbon future.

The transition to renewable energy is not just an environmental imperative; it is also an economic opportunity. The renewable energy sector is one of the fastest-growing industries globally, creating millions of jobs and driving innovation. From large-scale wind farms and solar power plants to community-owned microgrids and household solar systems, renewable energy projects are empowering communities and fostering economic development.

Structure of the Book

This book is organized into twenty chapters, each addressing a different aspect of renewable energy and its role in combating climate change. Here is a brief overview of what you can expect:

Chapter 1: Introduction to Renewable Energy

Lays the groundwork by explaining the various types of renewable energy and their significance.

Chapter 2: Historical Perspective

Traces the evolution of energy consumption and the shift towards renewable sources.

Chapter 3: The Science of Climate Change

Provides an in-depth look at the scientific principles behind climate change and the necessity of renewable energy.

Chapters 4-8

Explore individual renewable energy sources: solar, wind, hydro, geothermal, and biomass, detailing their technologies, benefits, and challenges.

Chapter 9: Energy Storage and Grid Integration

Discusses the importance of energy storage and smart grids in maximizing renewable energy use.

Chapter 10: Policy and Legislation

Reviews global policies and regulations that promote renewable energy adoption.

Chapter 11: Economic Impacts

Examines the economic benefits of renewable energy, including job creation and investment opportunities.

Chapter 12: Environmental and Social Benefits

Highlights the broader environmental and social advantages of renewable energy.

Chapter 13: Case Studies

Presents real-world examples of successful renewable energy projects.

Chapter 14: Technological Innovations

Discusses the latest advancements in renewable energy technologies.

Chapter 15: Barriers to Adoption

Identifies the obstacles to widespread renewable energy adoption and potential solutions.

Chapter 16: The Role of Corporations and Industry

Explores how businesses are integrating renewable energy into their operations.

Chapter 17: Community and Grassroots Movements

Examines the impact of local initiatives in driving renewable energy adoption.

Chapter 18: The Future of Renewable Energy

Looks ahead to the future potential and emerging trends in renewable energy.

Chapter 19: Renewable Energy and Developing Nations

Focuses on the unique challenges and opportunities for renewable energy in developing countries.

Chapter 20: Conclusion and Call to Action

Summarizes key points and urges readers to take action in supporting the renewable energy transition.

A Call to Action

The path to a sustainable future is clear, but it requires collective effort and commitment. Individuals, businesses, and governments all have a role to play in driving the renewable energy revolution. By making informed choices, advocating for supportive policies, and investing in clean technologies, we can accelerate the transition to a low-carbon economy.

This book aims to inform, inspire, and empower you to be part of this transformative journey. Together, we can combat climate change, protect our planet, and ensure a prosperous future for generations to come.

Welcome to "The Role of Renewable Energy in Combating Climate Change." Let us embark on this journey towards a sustainable and resilient future.

Chapter 1
Introduction to Renewable Energy

The Importance of Renewable Energy in Combating Climate Change

As the world grapples with the escalating crisis of climate change, renewable energy sources have emerged as the linchpin of global efforts to reduce greenhouse gas emissions and transition towards a sustainable future. Unlike fossil fuels, which release significant amounts of carbon dioxide and other harmful pollutants when burned, renewable energy sources generate power without depleting the Earth's natural resources or causing environmental degradation.

Fundamental Concepts of Renewable Energy

Renewable energy encompasses a diverse array of technologies and sources, each harnessing natural processes to generate power. The primary forms of renewable energy include solar, wind, hydro, geothermal, and biomass. Understanding these fundamental concepts is essential for appreciating their potential and limitations.

1: Solar Energy

How it Works: Solar energy harnesses the power of the sun through photovoltaic cells that convert sunlight directly into electricity or through solar thermal systems that capture and concentrate solar heat to generate power.

Applications: Solar panels are commonly used in residential, commercial, and utility-scale installations. Solar thermal systems can be used for water heating and power generation.

Advantages: Abundant, widely available, and capable of being deployed in diverse environments. Reduces dependence on fossil fuels and lowers greenhouse gas emissions.

Challenges: Intermittency issues due to weather and daylight variations. Requires significant initial investment and space for large-scale installations.

2: Wind Energy

How it Works: Wind energy is generated by converting the kinetic energy of wind into mechanical power using wind turbines. This mechanical power is then transformed into electricity.

Applications: Wind farms can be located onshore or offshore, with the latter offering higher wind speeds and more consistent energy generation.

Advantages: Clean, renewable, and capable of producing significant amounts of electricity. Wind farms can coexist with agricultural land.

Challenges: Wind variability and intermittency. Potential impacts on local wildlife and noise concerns for nearby communities.

3: Hydropower

How it Works: Hydropower captures the energy of flowing or falling water to generate electricity, typically using dams or run-of-river systems.

Applications: Large-scale hydroelectric dams, small-scale micro-hydro systems, and pumped storage for energy storage and grid stability.

Advantages: Reliable and consistent power generation. Provides a flexible source of energy that can be adjusted to meet demand.

Challenges: Environmental and social impacts of large dams, such as habitat disruption, displacement of communities, and changes in water flow.

4: Geothermal Energy

How it Works: Geothermal energy harnesses the heat from within the Earth's crust to generate electricity or provide direct heating. This can be done through geothermal power plants or ground-source heat pumps.

Applications: Electricity generation, district heating systems, and direct-use applications such as greenhouses and aquaculture.

Advantages: Stable and reliable energy source with low emissions. Provides baseload power, complementing intermittent renewable sources.

Challenges: Limited to regions with accessible geothermal resources. High initial costs for drilling and plant construction.

5: Biomass Energy

How it Works: Biomass energy is produced from organic materials, such as agricultural residues, forestry products, and organic waste. These materials can be burned directly for heat or converted into biofuels and biogas.

Applications: Heating, electricity generation, and transportation fuels. Biomass can also be used in combined heat and power (CHP) systems.

Advantages: Utilizes waste materials, reducing landfill use and methane emissions. Can provide a continuous source of energy.

Challenges: Emissions from combustion, although lower than fossil fuels. Land use concerns and competition with food production.

The Current State of Global Energy Consumption

The global energy landscape is currently dominated by fossil fuels, which account for approximately 80% of the world's energy consumption. Coal, oil, and natural gas are the primary sources, providing energy for electricity generation, transportation, heating, and industrial processes. Despite their abundance and established infrastructure, fossil fuels are major contributors to greenhouse gas emissions, air pollution, and environmental degradation.

In contrast, renewable energy sources have been steadily gaining ground, driven by technological advancements, decreasing costs, and supportive policies. In 2020, renewables accounted for nearly 29% of global electricity generation, with wind and solar leading the growth. However, their share of total energy consumption remains relatively low compared to fossil fuels, highlighting the need for accelerated adoption and integration.

Environmental Impacts of Fossil Fuels

The reliance on fossil fuels has profound environmental impacts, contributing significantly to climate change and ecological destruction. Key environmental issues associated with fossil fuels include:

1: Greenhouse Gas Emissions: The combustion of fossil fuels releases large amounts of carbon dioxide (CO_2) and other greenhouse gases, which trap heat in the atmosphere and drive global warming. This leads to rising temperatures, melting ice caps, sea-level rise, and more frequent and severe weather events.

2: Air Pollution: Burning fossil fuels produces pollutants such as sulfur dioxide (SO_2), nitrogen oxides (NO_x), particulate matter, and volatile organic compounds (VOCs). These pollutants contribute to smog, acid rain, respiratory illnesses, and cardiovascular diseases.

3: Water Pollution: Extraction and processing of fossil fuels can contaminate water sources with hazardous chemicals, heavy metals, and hydrocarbons. Oil spills, coal mining runoff, and hydraulic fracturing (fracking) fluids pose significant risks to aquatic ecosystems and drinking water supplies.

4: Habitat Destruction: Fossil fuel extraction often involves habitat destruction, deforestation, and land degradation.

This impacts biodiversity, disrupts ecosystems, and displaces wildlife and local communities.

5: Resource Depletion: Fossil fuels are finite resources, and their extraction depletes reserves that took millions of years to form. As easily accessible sources are exhausted, the industry increasingly turns to more environmentally damaging extraction methods.

Conclusion

Renewable energy represents a vital solution to the pressing challenges of climate change and environmental degradation. By harnessing natural, inexhaustible resources such as sunlight, wind, water, geothermal heat, and organic materials, renewable energy technologies offer a sustainable and cleaner alternative to fossil fuels. The transition to renewable energy is essential for reducing greenhouse gas emissions, improving air quality, and ensuring a stable and resilient energy future. As the world seeks to mitigate the impacts of climate change, the role of renewable energy cannot be overstated.

Chapter 2
Historical Perspective

The Early Days of Energy Consumption

Humanity's relationship with energy dates back to prehistoric times when early humans relied on simple, renewable sources of energy for survival. Fire, harnessed from burning wood and other organic materials, provided heat, light, and a means to cook food. This use of biomass was humanity's primary energy source for thousands of years.

The Agricultural Revolution

The Agricultural Revolution, which began around 10,000 BC, marked a significant shift in energy consumption. As societies transitioned from nomadic hunter-gatherer lifestyles to settled agricultural communities, the demand for energy increased. Humans began to harness animal power for plowing fields, transportation, and grinding grain. Wind and water power also started to play a role, with the use of windmills and watermills for grinding grain and pumping water.

The Dawn of the Industrial Revolution

The Industrial Revolution, beginning in the late 18th century, brought about profound changes in energy consumption and production. This era was characterized by the transition from agrarian economies to industrialized ones, driven by advancements in technology and mechanization. Key developments during this period include:

1: Coal as a Primary Energy Source: The discovery and exploitation of coal reserves provided a new, powerful source of energy. Coal-fired steam engines powered factories, trains, and ships, leading to unprecedented economic growth and urbanization.

2: The Steam Engine: Invented by James Watt in the 1760s, the steam engine revolutionized transportation and manufacturing. It enabled the efficient conversion of coal into mechanical energy, fueling the expansion of industries and infrastructure.

3: Textile Industry: The mechanization of textile production, exemplified by inventions such as the spinning jenny and power loom, transformed the industry. Factories sprang up, consuming vast amounts of coal to power their machinery.

4: Iron and Steel Production: Innovations in metallurgy, including the use of coke (a coal derivative) in blast furnaces, led to increased production of iron and steel. These materials were essential for building railways, bridges, and machinery.

The Rise of Oil and Electricity

The late 19th and early 20th centuries saw the rise of new energy sources and technologies that further transformed societies:

1: Oil: The drilling of the first commercial oil well in Titusville, Pennsylvania, in 1859 marked the beginning of the oil industry. Oil quickly became a critical energy source, particularly for transportation with the advent of the internal combustion engine and the automobile.

2: Electricity: The development of electrical power generation and distribution systems revolutionized energy consumption. Thomas Edison's invention of the practical incandescent light bulb and the establishment of power plants and electrical grids brought electricity into homes and businesses, improving quality of life and enabling new technological advancements.

The Mid-20th Century: The Nuclear Age

The mid-20th century introduced nuclear energy as a new, potent source of power. Nuclear reactors harnessed the energy released from nuclear fission to generate electricity. While initially promising as a clean energy source, concerns about safety, radioactive waste, and the potential for catastrophic accidents tempered the enthusiasm for nuclear power.

The Environmental Movement and the Shift Toward Renewables

By the late 20th century, growing awareness of the environmental impacts of fossil fuels sparked a global environmental movement. Key milestones in this shift include:

1: Environmental Legislation: In response to rising pollution levels and environmental degradation, governments began to enact environmental protection laws. Notable examples include the Clean Air Act and the Clean Water Act in the United States.

2: Energy Crises: The oil crises of the 1970s, characterized by oil embargoes and skyrocketing prices, underscored the vulnerabilities of fossil fuel dependence. These crises prompted investments in energy efficiency and alternative energy sources.

3: Scientific Consensus on Climate Change: By the late 20th century, scientific evidence increasingly linked fossil fuel combustion to global climate change. Reports from the Intergovernmental Panel on Climate Change (IPCC) highlighted the urgent need to reduce greenhouse gas emissions.

4: Technological Advancements: Advancements in renewable energy technologies, such as photovoltaic cells, wind turbines, and energy storage systems, made renewables more viable and cost-effective. Governments and private sectors began investing in research and development to improve efficiency and reduce costs.

The 21st Century: A Renewable Revolution

The 21st century has witnessed significant strides toward renewable energy adoption, driven by technological innovation, policy support, and growing public awareness of climate change:

1: Cost Reductions: The costs of renewable energy technologies have plummeted over the past two decades. Solar and wind power have become some of the cheapest sources of electricity in many regions, competing favorably with fossil fuels.

2: Global Agreements: International agreements, such as the Paris Agreement of 2015, have committed countries to reducing greenhouse gas emissions and increasing the use of renewable energy. These agreements have spurred national policies and investments in clean energy.

3: Corporate and Financial Sector Engagement: Many corporations have set ambitious renewable energy targets, driven by sustainability goals and economic considerations. The financial sector has also increasingly recognized the risks of climate change, leading to more investments in renewable energy projects.

4: Technological Innovations: Breakthroughs in energy storage, such as advanced batteries, and smart grid technologies have addressed some of the intermittency challenges of renewables. Innovations in offshore wind, floating solar panels, and green hydrogen production are expanding the possibilities for renewable energy deployment.

5: Community and Local Initiatives: Grassroots movements and community-led renewable energy projects have demonstrated the potential for local solutions to contribute significantly to the energy transition. Cooperative models and decentralized energy systems empower communities and enhance energy resilience.

Conclusion

The historical perspective on energy consumption reveals a dynamic evolution influenced by technological advancements, societal needs, and environmental considerations. From the early reliance on biomass to the dominance of fossil fuels during the Industrial Revolution, and the ongoing shift toward renewable energy in the 21st century, humanity's energy choices have shaped and been shaped by broader economic, social, and environmental factors. As we confront the challenges of climate change, the lessons of history underscore the importance of innovation, policy support, and collective action in transitioning to a sustainable energy future.

Chapter 3
The Science of Climate Change

Introduction to Climate Change

Climate change refers to significant changes in global temperatures and weather patterns over time. While climate change is a natural phenomenon, current trends are largely driven by human activities, particularly the burning of fossil fuels, deforestation, and industrial processes. Understanding the science behind climate change is essential for recognizing the urgency of the issue and the critical role of renewable energy in mitigating its impacts.

The Greenhouse Effect

The greenhouse effect is a natural process that warms the Earth's surface. When the Sun's energy reaches the Earth, some of it is reflected back to space, and the rest is absorbed and re-radiated as heat. Greenhouse gases in the atmosphere, including carbon dioxide (CO_2), methane (CH_4), and nitrous oxide (N_2O), trap this heat, preventing it from escaping into space and thus warming the planet.

1: Natural Greenhouse Effect: The natural greenhouse effect is essential for life on Earth, maintaining temperatures that allow ecosystems and human societies to thrive. Without it, the planet's average temperature would be about -18°C (0°F), rather than the current 15°C (59°F).

2: Enhanced Greenhouse Effect: Human activities have increased the concentration of greenhouse gases in the atmosphere, enhancing the natural greenhouse effect and causing more heat to be trapped. This leads to global warming and climate change.

Carbon Emissions

Carbon emissions, particularly carbon dioxide (CO_2), are the primary driver of the enhanced greenhouse effect. Key sources of carbon emissions include:

1: Fossil Fuels: The burning of coal, oil, and natural gas for electricity, heat, and transportation is the largest source of CO_2 emissions. These activities release carbon that has been stored in fossil fuels for millions of years, significantly altering the carbon cycle.

2: Deforestation: Trees and forests act as carbon sinks, absorbing CO_2 from the atmosphere through photosynthesis. When forests are cleared for agriculture, urban development, or logging, this stored carbon is released back into the atmosphere.

3: Industrial Processes: Various industrial activities, such as cement production, steelmaking, and chemical manufacturing, release CO_2 and other greenhouse gases as byproducts.

The Role of Human Activities

Human activities have significantly altered the Earth's natural systems, leading to an increase in greenhouse gas concentrations and contributing to climate change. Key human activities impacting the climate include:

1: Energy Production: The burning of fossil fuels for energy is the largest source of greenhouse gas emissions. Power plants, industrial facilities, and vehicles all contribute to this problem.

2: Land Use Changes: Agriculture, deforestation, and urbanization have transformed landscapes, reducing the Earth's capacity to absorb CO_2 and increasing emissions from soil and vegetation.

3: Agriculture: Agricultural practices, including livestock farming and rice cultivation, produce methane and nitrous oxide, potent greenhouse gases. The use of synthetic fertilizers also contributes to nitrous oxide emissions.

4: Waste Management: Landfills and waste treatment processes produce methane as organic waste decomposes. Inefficient waste management practices exacerbate this issue.

Impacts of Climate Change

The impacts of climate change are already being felt around the world and are projected to intensify in the coming decades. Key impacts include:

1: Rising Temperatures: Global temperatures have increased by approximately 1.2°C (2.2°F) since the late 19th century. This warming trend leads to more frequent and intense heatwaves.

2: Sea Level Rise: Melting glaciers and ice sheets, combined with the thermal expansion of seawater, are causing sea levels to rise. This threatens coastal communities, infrastructure, and ecosystems.

3: Extreme Weather Events: Climate change is linked to an increase in the frequency and severity of extreme weather events, including hurricanes, droughts, floods, and wildfires.

4: Ocean Acidification: The absorption of excess CO_2 by the oceans is causing acidification, which harms marine life, particularly organisms with calcium carbonate shells or skeletons.

5: Ecosystem Disruption: Changes in temperature and precipitation patterns disrupt ecosystems, affecting biodiversity, species distribution, and the timing of natural events such as migration and flowering.

6: Human Health: Climate change poses direct and indirect threats to human health, including heat-related illnesses, respiratory issues from air pollution, and the spread of vector-borne diseases.

The Role of Renewable Energy in Mitigating Climate Change

Renewable energy plays a crucial role in mitigating climate change by reducing greenhouse gas emissions and providing a sustainable alternative to fossil fuels. Key benefits of renewable energy include:

1: Reducing Carbon Emissions: Renewable energy sources, such as solar, wind, hydro, and geothermal, produce little to no greenhouse gas emissions during operation. By replacing fossil fuels with renewables, we can significantly reduce CO_2 emissions.

2: Energy Efficiency: Many renewable energy technologies are more efficient than traditional fossil fuel-based systems. For example, electric vehicles powered by renewable electricity are more efficient than internal combustion engine vehicles.

3: Energy Independence: Renewable energy sources are often locally available, reducing dependence on imported fossil fuels and enhancing energy security.

4: Economic Benefits: The renewable energy sector creates jobs and stimulates economic growth. Investments in renewable energy infrastructure can drive technological innovation and increase competitiveness.

5: Environmental and Health Benefits: Reducing reliance on fossil fuels improves air and water quality, reducing health risks associated with pollution. Protecting ecosystems from the impacts of fossil fuel extraction and combustion also benefits biodiversity and natural habitats.

Transitioning to a Renewable Energy Future

The transition to a renewable energy future requires coordinated efforts across multiple sectors, including government, industry, and civil society. Key strategies for accelerating the adoption of renewable energy include:

1: Policy Support: Governments can implement policies and incentives to promote renewable energy, such as subsidies, tax credits, feed-in tariffs, and renewable energy standards.

2: Investment in Research and Development: Continued investment in research and development is essential for advancing renewable energy technologies, improving efficiency, and reducing costs.

3: Infrastructure Development: Expanding and modernizing energy infrastructure, including grids and storage systems, is crucial for integrating renewable energy into the existing energy system.

4: Public Awareness and Education: Increasing public awareness and understanding of climate change and renewable energy can drive consumer demand and support for clean energy solutions.

5: International Cooperation: Global challenges require global solutions. International cooperation and knowledge sharing can accelerate the transition to renewable energy and address climate change more effectively.

Conclusion

The science of climate change underscores the urgent need to reduce greenhouse gas emissions and transition to renewable energy sources. Human activities have significantly contributed to the enhanced greenhouse effect, leading to global warming and a host of environmental and social impacts. Renewable energy offers a sustainable and effective solution to mitigate climate change, providing a path toward a cleaner, healthier, and more resilient future. By understanding the science behind climate change and the critical role of renewable energy, we can take informed actions to protect our planet and ensure a sustainable future for generations to come.

Chapter 4
Solar Power
Harnessing the Sun

Introduction to Solar Energy

Solar energy, the most abundant renewable energy source available, harnesses the power of the Sun to generate electricity and heat. The Sun delivers an immense amount of energy to the Earth every day—far more than the world's current energy consumption. As technology advances and costs decrease, solar energy has become a critical component in the global transition to sustainable energy. This chapter explores the different technologies used to harness solar power, their applications, and the latest advancements driving the industry forward.

Photovoltaic (PV) Cells

Photovoltaic cells, commonly known as solar cells, are the most widely recognized technology for converting sunlight directly into electricity.

1: How Photovoltaic Cells Work

Basic Principle: Photovoltaic cells operate on the principle of the photovoltaic effect. When sunlight strikes the surface of a solar cell, it excites electrons in the cell's semiconductor material, creating an electric current.

Materials: The most common semiconductor material used in PV cells is silicon, although other materials like cadmium telluride (CdTe) and copper indium gallium selenide (CIGS) are also used.

Structure: A typical PV cell consists of a thin semiconductor layer sandwiched between two conductive layers. When sunlight hits the semiconductor, it generates free electrons that move toward the front conductive layer, creating a flow of electric current.

2: Types of Photovoltaic Systems

Monocrystalline Silicon Cells: Made from a single crystal structure, these cells offer high efficiency and durability. They are typically more expensive than other types of cells.

Polycrystalline Silicon Cells: Made from multiple silicon crystals, these cells are less efficient than monocrystalline cells but are cheaper to produce.

Thin-Film Solar Cells: These cells are made by depositing one or more layers of photovoltaic material onto a substrate. They are flexible, lightweight, and can be made from various materials, including amorphous silicon, CdTe, and CIGS.

3: Applications of Photovoltaic Systems

Residential: PV systems can be installed on rooftops to provide electricity for homes. They can be connected to the grid (grid-tied) or operate independently with battery storage (off-grid).

Commercial: Businesses and industrial facilities can install larger PV systems to reduce energy costs and improve sustainability.

Utility-Scale: Large-scale solar farms generate significant amounts of electricity, feeding it directly into the grid. These installations can cover vast areas and produce hundreds of megawatts of power.

Solar Thermal Systems

Solar thermal systems capture and concentrate sunlight to generate heat, which can then be used for various applications, including electricity generation, water heating, and space heating.

1: Types of Solar Thermal Systems

Solar Water Heaters: These systems use solar collectors to absorb sunlight and heat water for domestic or commercial use. They can be active (using pumps) or passive (relying on natural convection).

Solar Space Heating: Similar to solar water heaters, these systems use solar collectors to heat air or water for space heating in buildings.

Concentrated Solar Power (CSP): CSP systems use mirrors or lenses to focus sunlight onto a small area, generating intense heat that can be used to produce steam and drive a turbine to generate electricity. CSP technologies include parabolic troughs, solar power towers, and dish Stirling systems.

2: Advantages of Solar Thermal Systems

Efficiency: Solar thermal systems are highly efficient at converting sunlight into heat.

Cost-Effectiveness: For applications like water heating, solar thermal systems can be more cost-effective than PV systems.

Energy Storage: CSP systems can store thermal energy in molten salts, allowing for electricity generation even when the sun isn't shining.

Latest Advancements in Solar Technology

The solar industry is continually evolving, with ongoing research and development leading to new technologies and improvements in efficiency, cost, and applicability.

1: Bifacial Solar Panels

Design: Bifacial solar panels capture sunlight on both sides, increasing energy yield. They can be mounted vertically or at an angle to maximize exposure.

Applications: Particularly effective in environments with high albedo surfaces (e.g., snow, sand, or white rooftops) that reflect sunlight onto the backside of the panel.

2: Perovskite Solar Cells

Material: Perovskite materials have shown great promise due to their high efficiency and low production costs. They can be manufactured using simple, low-temperature processes.

Challenges: Stability and durability remain issues, but ongoing research aims to overcome these hurdles.

3: Building-Integrated Photovoltaics (BIPV)

Integration: BIPV integrates PV materials into building elements like windows, facades, and roofs. This approach not only generates electricity but also enhances building aesthetics and reduces material costs.

Applications: Suitable for new constructions and retrofits, BIPV can help buildings achieve net-zero energy status.

4: Floating Solar Farms

Concept: Floating solar farms are installed on bodies of water, such as lakes, reservoirs, and coastal areas. This approach saves land space and can improve PV efficiency due to the cooling effect of water.

Benefits: Reduces water evaporation from reservoirs and can be combined with aquaculture.

5: Solar Tracking Systems

Mechanism: Solar tracking systems move solar panels to follow the sun's path, maximizing exposure and energy generation throughout the day.

Types: Single-axis trackers move panels along one axis (typically east-west), while dual-axis trackers adjust both azimuth and altitude.

6: Advanced Energy Storage

Batteries: Improved battery technologies, such as lithium-ion and emerging solid-state batteries, enhance the storage of solar energy, making it available when sunlight isn't.

Thermal Storage: Innovations in thermal storage materials and systems improve the efficiency and practicality of storing solar thermal energy.

The Potential of Solar Energy

Solar energy has the potential to meet a significant portion of global energy needs. Its advantages include:

1: Abundance: The Sun provides an almost limitless supply of energy. The amount of solar energy that reaches the Earth in one hour is more than the total global energy consumption for a year.

2: Sustainability: Solar energy is renewable and does not deplete natural resources or cause environmental harm during operation.

3: Scalability: Solar technologies can be deployed at various scales, from small residential systems to large utility-scale solar farms.

4: Decentralization: Solar energy allows for decentralized power generation, reducing reliance on centralized power plants and enhancing energy security.

Challenges and Solutions

While solar energy offers immense potential, several challenges must be addressed to maximize its adoption and effectiveness:

1: Intermittency: Solar power generation is dependent on sunlight, which varies with weather and time of day.

Solution: Combining solar with energy storage and other renewable sources, as well as using grid management technologies, can mitigate intermittency issues.

2: Initial Costs: The upfront costs of solar installations can be high, although they have been decreasing.

Solution: Financial incentives, subsidies, and innovative financing models can help lower barriers to adoption.

3: Land Use: Large solar farms require significant land areas, which can compete with other land uses.

Solution: Utilizing rooftops, integrating solar into urban environments, and developing floating solar farms can optimize land use.

4: Material and Manufacturing: The production of solar panels involves energy-intensive processes and the use of certain raw materials.

Solution: Advances in manufacturing efficiency, recycling of materials, and development of sustainable alternatives can address these concerns.

Conclusion

Solar energy represents a cornerstone of the global transition to renewable energy. Through photovoltaic and solar thermal technologies, we can harness the Sun's vast energy potential to generate electricity and heat sustainably. Continuous advancements in solar technology, combined with supportive policies and innovative solutions, are driving the growth of solar energy worldwide.

As we strive to combat climate change and reduce our dependence on fossil fuels, solar power stands out as a key player in creating a cleaner, more resilient energy future.

Chapter 5
Wind Energy
Power from the Air

Introduction to Wind Energy

Wind energy harnesses the power of moving air to generate electricity. As one of the fastest-growing renewable energy sources, wind energy plays a crucial role in the global shift towards sustainable and clean energy. This chapter provides an in-depth look at the mechanics of wind turbines, the development of onshore and offshore wind farms, the future potential of wind power, and the challenges and benefits associated with this renewable energy source.

Mechanics of Wind Turbines

Wind turbines convert the kinetic energy of wind into mechanical energy, which is then transformed into electrical energy.

1: Basic Components of Wind Turbines

Rotor Blades: The rotor consists of blades (usually three) that capture wind energy. The blades are designed to rotate when wind passes over them, creating lift and causing the rotor to spin.

Nacelle: The nacelle sits atop the tower and houses key components, including the gearbox, generator, and control systems. It also supports the rotor hub.

Tower: The tower elevates the rotor and nacelle to a height where they can capture stronger and more consistent winds. Towers can be made from steel, concrete, or hybrid materials.

Generator: Located inside the nacelle, the generator converts the mechanical energy from the rotor into electrical energy.

Gearbox: The gearbox increases the rotational speed from the rotor to a level suitable for generating electricity. Some modern turbines use direct-drive systems that eliminate the need for a gearbox.

Yaw System: This system rotates the nacelle and rotor to face the wind, optimizing energy capture.

Control System: The control system monitors and adjusts turbine operations to ensure optimal performance and safety.

2: Types of Wind Turbines

Horizontal-Axis Wind Turbines (HAWT): The most common type, with a horizontal rotor shaft and blades that rotate around a horizontal axis. These turbines are typically large and used in both onshore and offshore wind farms.

Vertical-Axis Wind Turbines (VAWT): These turbines have a vertical rotor shaft and blades that rotate around a vertical axis. They are less common but can be useful in specific applications, such as urban environments.

3: Wind Turbine Operation

Wind Capture: As wind flows over the blades, it creates a pressure difference that generates lift, causing the rotor to spin.

Energy Conversion: The rotor's mechanical energy is transferred through the gearbox to the generator, which converts it into electrical energy.

Electricity Transmission: The generated electricity is transmitted through cables down the tower to the electrical grid or storage systems.

Onshore and Offshore Wind Farms

Wind farms consist of multiple wind turbines connected to form a single power-generating system. They can be located on land (onshore) or in bodies of water (offshore).

1: Onshore Wind Farms

Location: Onshore wind farms are situated on land, typically in areas with strong and consistent winds, such as plains, ridges, and coastal regions.

Advantages: Onshore wind farms are generally cheaper to build and maintain than offshore wind farms. They can be integrated with existing infrastructure and offer easier access for maintenance.

Challenges: Onshore wind farms can face issues related to land use, noise, and visual impact. They may also encounter opposition from local communities concerned about aesthetics and wildlife impacts.

2: Offshore Wind Farms

Location: Offshore wind farms are located in bodies of water, such as oceans, seas, and large lakes. They are typically situated where wind speeds are higher and more consistent.

Advantages: Offshore wind farms can generate more energy due to stronger and steadier winds. They have less visual and noise impact on communities and can be built on a larger scale.

Challenges: Offshore wind farms are more expensive to build and maintain due to the harsh marine environment. They require advanced technology and specialized vessels for installation and maintenance.

3: Floating Wind Farms

Emerging Technology: Floating wind farms use turbines mounted on floating platforms anchored to the seabed. This technology allows for deployment in deeper waters where traditional fixed-bottom turbines are not feasible.

Potential: Floating wind farms can access vast offshore wind resources, expanding the potential for wind energy generation.

Future Potential of Wind Power

Wind energy has significant potential to contribute to the global energy mix and mitigate climate change.

1: Technological Advancements

Larger Turbines: Advances in turbine design and materials are leading to larger and more efficient turbines, capable of generating more power from the same amount of wind.

Improved Materials: The use of advanced materials, such as carbon fiber and composite materials, enhances turbine performance and durability.

Smart Turbines: Integration of sensors, artificial intelligence, and data analytics improves turbine efficiency, predictive maintenance, and grid integration.

2: Grid Integration and Storage

Energy Storage: Developing efficient energy storage solutions, such as batteries and pumped hydro storage, helps manage the intermittent nature of wind power and ensures a stable energy supply.

Grid Upgrades: Modernizing electrical grids to handle distributed and variable energy sources is crucial for maximizing the potential of wind energy.

3: Policy and Market Support

Incentives and Subsidies: Government policies, incentives, and subsidies play a critical role in promoting wind energy development and investment.

Corporate Demand: Growing corporate demand for renewable energy, driven by sustainability goals and carbon reduction commitments, is boosting wind energy deployment.

Challenges of Wind Energy

While wind energy offers numerous benefits, it also faces several challenges that need to be addressed:

1: Intermittency and Reliability

Challenge: Wind energy production is variable and dependent on wind availability, leading to intermittency issues.

Solution: Combining wind energy with other renewable sources and energy storage can enhance reliability and stability.

2: Environmental Impact

Challenge: Wind farms can impact wildlife, particularly birds and bats, and alter local ecosystems.

Solution: Conducting thorough environmental assessments, implementing mitigation measures, and choosing appropriate locations can minimize environmental impacts.

3: Land Use and Aesthetics

Challenge: Onshore wind farms require significant land and can face opposition due to visual and noise impacts.

Solution: Engaging with local communities, optimizing turbine placement, and developing offshore and floating wind farms can address land use and aesthetic concerns.

4: Cost and Financing

Challenge: Initial costs for wind farm development and maintenance can be high, particularly for offshore projects.

Solution: Continued technological advancements, economies of scale, and supportive policies can reduce costs and attract investment.

Benefits of Wind Energy

Despite the challenges, wind energy offers numerous benefits that make it a vital component of a sustainable energy future:

1: Low Carbon Footprint

Wind energy produces no greenhouse gas emissions during operation, significantly reducing the carbon footprint compared to fossil fuels.

2: Renewable and Abundant

Wind is an inexhaustible natural resource, making wind energy a sustainable and long-term solution for energy generation.

3: Energy Independence

Wind energy enhances energy security by diversifying the energy mix and reducing reliance on imported fossil fuels.

4: Economic Growth

The wind energy industry creates jobs in manufacturing, installation, maintenance, and research, contributing to economic development.

5: Scalability and Flexibility

Wind energy systems can be deployed at various scales, from small community projects to large utility-scale wind farms, providing flexibility in energy planning.

Conclusion

Wind energy harnesses the power of the air to generate clean, renewable electricity, playing a critical role in combating climate change and achieving sustainable energy goals. Through advancements in wind turbine technology, the development of onshore and offshore wind farms, and the potential of emerging technologies like floating wind farms, wind energy continues to expand its contribution to the global energy mix. While challenges such as intermittency, environmental impacts, and costs need to be addressed, the benefits of wind energy—its low carbon footprint, abundance, energy independence, and economic potential—make it a cornerstone of a sustainable energy future. By understanding the mechanics, applications, and future potential of wind power, we can harness this powerful resource to build a cleaner, more resilient world.

Chapter 6
Hydropower
Energy from Water

Introduction to Hydropower

Hydropower, the generation of electricity using the energy of moving water, is one of the oldest and most established forms of renewable energy. It accounts for a significant portion of the world's renewable energy production, providing reliable and consistent power. This chapter explores the various types of hydropower, including large-scale dams and micro-hydro projects, and examines their environmental and social impacts, as well as their role in the renewable energy mix.

Types of Hydropower

Hydropower projects can vary greatly in size and scope, from massive dams that provide electricity to millions, to small-scale systems that power remote communities.

1: Large-Scale Hydropower

Dams and Reservoirs: The most common form of large-scale hydropower involves the construction of a dam on a river, creating a reservoir. Water released from the reservoir flows through turbines, generating electricity. Notable examples include the Three Gorges Dam in China and the Hoover Dam in the United States.

Run-of-the-River Systems: Unlike dam-based systems, run-of-the-river hydropower projects divert a portion of river flow through a canal or penstock to generate electricity without significantly altering the river's flow or requiring large reservoirs. These systems have a smaller environmental footprint compared to traditional dam projects.

2: Small-Scale and Micro-Hydropower

Small Hydropower: Projects with an installed capacity of up to 10 megawatts (MW) are considered small hydropower. These systems often serve small communities or industrial facilities and can be integrated into existing infrastructure, such as irrigation canals.

Micro-Hydropower: Systems with a capacity of less than 100 kilowatts (kW) are classified as micro-hydropower. They are typically used for off-grid applications, providing electricity to remote or rural areas where grid access is limited. These systems can be installed on small streams or rivers with minimal environmental impact.

3: Pumped Storage Hydropower

Mechanism: Pumped storage hydropower (PSH) involves two water reservoirs at different elevations. During periods of low electricity demand, water is pumped from the lower reservoir to the upper reservoir using excess grid electricity. During high demand, water is released from the upper reservoir to the lower reservoir, generating electricity through turbines.

Advantages: PSH acts as an energy storage system, helping to balance grid demand and supply, and supporting the integration of other variable renewable energy sources like wind and solar.

Environmental and Social Impacts of Hydropower

While hydropower is a clean and renewable energy source, it can have significant environmental and social impacts that need to be carefully managed.

1: Environmental Impacts

Ecosystem Disruption: Large dams and reservoirs can significantly alter local ecosystems, affecting fish populations, river flow, and water quality. Fish migration can be disrupted, leading to declines in certain species.

Habitat Loss: The creation of reservoirs can flood large areas, leading to the loss of terrestrial and aquatic habitats, and impacting biodiversity.

Greenhouse Gas Emissions: Although hydropower produces no direct emissions, reservoirs in tropical regions can emit methane, a potent greenhouse gas, due to the decomposition of organic matter underwater.

2: Social Impacts

Displacement: Large dam projects often require the relocation of communities living in the area to be flooded, leading to social and economic disruption for affected populations.

Cultural Heritage: The flooding of areas for reservoirs can submerge cultural and historical sites, leading to the loss of cultural heritage.

Livelihoods: Changes in river flow and water availability can impact agriculture, fishing, and other livelihoods dependent on water resources.

3: Mitigation Measures

Environmental Mitigation: Measures such as fish ladders, bypass systems, and environmental flow releases can help mitigate the impact on aquatic ecosystems and fish populations.

Social Mitigation: Adequate compensation, resettlement plans, and community engagement are essential to address the social impacts of displacement and ensure the well-being of affected communities.

The Role of Hydropower in the Renewable Energy Mix

Hydropower plays a crucial role in the global transition to renewable energy, offering several unique advantages.

1: Reliable and Consistent Power

Base Load Power: Unlike some other renewable energy sources, hydropower can provide a constant and reliable source of electricity, making it suitable for base load power generation.

Peak Load and Flexibility: Hydropower plants, especially pumped storage systems, can quickly adjust output to match demand, providing valuable flexibility and peak load capacity to the grid.

2: Energy Storage and Grid Stability

Pumped Storage: As the most widely used form of large-scale energy storage, pumped storage hydropower helps stabilize the grid, balancing supply and demand, and supporting the integration of intermittent renewables like wind and solar.

Grid Services: Hydropower plants can provide essential grid services such as frequency regulation, voltage support, and spinning reserve, contributing to overall grid stability and reliability.

3: Sustainable Development Goals

Renewable Energy Targets: Hydropower contributes significantly to achieving renewable energy targets set by various countries and international agreements aimed at reducing greenhouse gas emissions and combating climate change.

Economic Development: Hydropower projects can stimulate local economies, create jobs, and improve infrastructure, contributing to broader sustainable development goals.

Future Potential and Innovations in Hydropower

Advancements in technology and innovative approaches are expanding the potential of hydropower to contribute to a sustainable energy future.

1: Technological Innovations

Advanced Turbines: New turbine designs, such as fish-friendly and more efficient turbines, are being developed to reduce environmental impacts and improve performance.

Digitalization and Smart Grids: The integration of digital technologies, such as sensors and data analytics, enhances the operation and maintenance of hydropower plants, improving efficiency and reducing downtime.

2: Hybrid Systems

Solar-Hydro Hybrid: Combining hydropower with solar energy in hybrid systems can maximize the use of available resources, particularly in regions with variable water flow and abundant sunlight.

Wind-Hydro Hybrid: Integrating wind and hydropower can help balance the intermittency of wind energy, utilizing hydropower's flexibility to provide consistent power supply.

3: Sustainable Small-Scale Hydropower

Community-Based Projects: Small-scale and micro-hydropower projects can be developed and managed by local communities, providing reliable and sustainable energy while minimizing environmental impacts.

Run-of-the-River Expansion: Expanding the use of run-of-the-river systems offers a low-impact alternative to large dams, providing renewable energy without significant alteration of river ecosystems.

Conclusion

Hydropower, with its diverse range of applications from large-scale dams to micro-hydro projects, is a vital component of the renewable energy landscape. While it offers significant benefits in terms of reliable power generation, energy storage, and grid stability, it also presents environmental and social challenges that must be carefully managed. Through technological innovations, sustainable practices, and thoughtful integration with other renewable energy sources, hydropower can continue to play a critical role in the global transition to a sustainable energy future. By understanding the complexities and potential of hydropower, we can harness the power of water to build a cleaner, more resilient world.

Chapter 7
Geothermal Energy
Heat from the Earth

Introduction to Geothermal Energy

Geothermal energy, derived from the natural heat of the Earth's interior, is a sustainable and reliable source of power. It is harnessed by tapping into underground reservoirs of steam and hot water, which can be used directly for heating or to generate electricity. This chapter provides an overview of geothermal energy, its harnessing methods, and various applications, and discusses its potential in providing a stable and reliable energy source.

How Geothermal Energy is Harnessed

Geothermal energy harnessing involves extracting heat from beneath the Earth's surface. The process varies based on the type and temperature of the geothermal resource.

1: Types of Geothermal Resources

High-Temperature Resources: Found in regions with volcanic activity, such as Iceland and the Philippines, these resources are typically above 150°C and are used for electricity generation.

Medium-Temperature Resources: These resources, with temperatures between 90°C and 150°C, are often used for both electricity generation and direct heating applications.

Low-Temperature Resources: Below 90°C, these resources are mainly used for direct heating applications, such as district heating and agricultural uses.

Enhanced Geothermal Systems (EGS): These systems enhance or create geothermal resources by injecting water into hot dry rocks to produce steam for power generation.

2: Geothermal Power Plants

Dry Steam Plants: The oldest type of geothermal power plant, dry steam plants use steam directly from a geothermal reservoir to turn turbines and generate electricity. These plants are typically found in regions with very high-temperature resources.

Flash Steam Plants: The most common type, flash steam plants use high-pressure hot water from the reservoir. When the pressure is reduced, some of the water flashes into steam, which is then used to drive turbines.

Binary Cycle Plants: These plants use moderate-temperature geothermal resources. The geothermal water heats a secondary fluid with a lower boiling point in a heat exchanger. The secondary fluid vaporizes and drives the turbines. Binary cycle plants do not emit geothermal gases, making them environmentally friendly.

3: Direct Use Applications

District Heating Systems: Geothermal heat is used to warm entire communities by circulating hot water through a network of pipes.

Agricultural Uses: Geothermal heat is used for greenhouse heating, aquaculture, and soil warming to extend growing seasons and increase yields.

Industrial Uses: Geothermal energy is used in various industrial processes, such as drying crops, pasteurizing milk, and drying lumber.

Geothermal Heat Pumps (GHPs): These systems use the stable ground temperature a few meters below the surface to provide heating and cooling for buildings. GHPs are efficient and widely used in residential and commercial buildings.

Applications of Geothermal Energy

Geothermal energy has a broad range of applications, from electricity generation to direct heating, offering a versatile and stable energy source.

1: Electricity Generation

Global Distribution: Geothermal power plants are found in over 20 countries, including the United States, Iceland, Indonesia, and New Zealand. These plants provide a significant portion of the electricity needs in some regions.

Base Load Power: Unlike wind and solar energy, geothermal energy provides continuous base load power, as it is not dependent on weather conditions. This makes it a reliable source of electricity.

2: Direct Heating

District Heating: Cities like Reykjavik, Iceland, use geothermal energy for district heating, significantly reducing their reliance on fossil fuels.

Agricultural Applications: In countries like the Netherlands, geothermal energy is used to heat greenhouses, allowing for year-round production of fruits and vegetables.

Industrial Processes: Industries in countries like Kenya use geothermal energy for processes such as drying tea and coffee, which require significant heat.

3: Geothermal Heat Pumps

Residential Use: Geothermal heat pumps are used in homes across Europe, North America, and Asia for space heating and cooling, providing a highly efficient alternative to traditional HVAC systems.

Commercial Use: Large buildings and facilities, such as schools, hospitals, and office buildings, utilize geothermal heat pumps to achieve significant energy savings and reduce greenhouse gas emissions.

Potential of Geothermal Energy

Geothermal energy offers immense potential as a stable and reliable energy source, with several key advantages over other renewables.

1: Renewable and Sustainable

Inexhaustible Source: Geothermal energy is derived from the Earth's internal heat, which is virtually inexhaustible over human time scales.

Low Emissions: Geothermal power plants emit minimal greenhouse gases compared to fossil fuel plants, contributing significantly to reducing carbon footprints.

2: Reliable and Consistent

Base Load Capability: Geothermal power provides a constant and stable supply of electricity, unlike intermittent sources like wind and solar. This reliability makes it a valuable component of a diversified energy mix.

High Capacity Factor: Geothermal plants have a high capacity factor, often exceeding 90%, meaning they produce electricity almost continuously throughout the year.

3: Scalability and Flexibility

Variety of Applications: Geothermal energy can be used for a wide range of applications, from large-scale power generation to small-scale direct heating projects.

Localized Use: Geothermal projects can be developed locally, providing energy independence and security for communities and reducing the need for energy imports.

4: Technological Advancements

Enhanced Geothermal Systems (EGS): Advances in EGS technology are expanding the potential for geothermal energy by enabling the extraction of heat from dry and impermeable rocks.

Drilling and Exploration: Innovations in drilling techniques and seismic imaging are improving the efficiency and feasibility of geothermal energy extraction, reducing costs and expanding accessible resources.

Challenges and Solutions in Geothermal Energy

While geothermal energy offers numerous benefits, it also faces challenges that need to be addressed for its widespread adoption.

1: Resource Location and Accessibility

Challenge: Geothermal resources are not uniformly distributed and are often located in geologically active regions, limiting their availability.

Solution: Enhanced Geothermal Systems (EGS) and advanced exploration techniques can expand the range of accessible geothermal resources, making them viable in more locations.

2: High Initial Costs

Challenge: The initial costs of drilling and plant construction can be high, posing a barrier to investment.

Solution: Government incentives, subsidies, and innovative financing models can help offset initial costs and attract investment. Technological advancements and economies of scale are also reducing costs over time.

3: Environmental Concerns

Challenge: Geothermal projects can have localized environmental impacts, such as land subsidence and the release of harmful gases from underground.

Solution: Careful site selection, environmental monitoring, and mitigation measures, such as reinjecting geothermal fluids back into the reservoir, can minimize environmental impacts.

4: Public Perception and Awareness

Challenge: Lack of public awareness and understanding of geothermal energy can hinder its acceptance and development.

Solution: Education campaigns, community engagement, and transparent communication about the benefits and impacts of geothermal energy can improve public perception and support.

Conclusion

Geothermal energy, with its diverse applications and reliable nature, holds significant promise as a key component of the global renewable energy portfolio. From large-scale power generation to direct heating and cooling, geothermal energy offers a sustainable and low-emission alternative to fossil fuels. While challenges such as high initial costs and resource accessibility need to be addressed, advancements in technology and supportive policies are paving the way for greater adoption and expansion of geothermal energy.

By harnessing the Earth's natural heat, we can move towards a more sustainable and resilient energy future, reducing our reliance on fossil fuels and mitigating the impacts of climate change.

Chapter 8
Biomass
Organic Energy

Introduction to Biomass Energy

Biomass energy, derived from organic materials, is one of humanity's oldest sources of energy. As a renewable resource, biomass can be transformed into electricity, heat, and biofuels, playing a significant role in reducing reliance on fossil fuels and mitigating climate change. This chapter explores the various sources of biomass, the processes for converting biomass into energy, and the sustainability of biomass energy.

Sources of Biomass

Biomass can be sourced from a variety of organic materials, making it a versatile and widely available energy resource.

1: Agricultural Residues

Crop Residues: Leftover parts of crops, such as straw, husks, and stalks, can be used as biomass. These residues are abundant and often underutilized.

Animal Manure: Livestock manure is another source of biomass, which can be processed through anaerobic digestion to produce biogas.

Energy Crops: Certain crops, such as switchgrass, miscanthus, and fast-growing trees like poplar and willow, are specifically cultivated for energy production due to their high biomass yield.

2: Forestry Products

Logging Residues: Wood chips, branches, and sawdust generated from logging and timber processing can be used as biomass.

Forest Thinnings: Managing forests for health and fire prevention often involves thinning operations, where smaller trees and underbrush are removed and can be used as biomass.

Wood Pellets: Manufactured from compressed sawdust and other wood residues, wood pellets are a popular form of biomass for heating and electricity generation.

3: Organic Waste

Municipal Solid Waste (MSW): Organic components of MSW, such as food scraps, yard waste, and paper, can be separated and processed to produce energy.

Industrial Waste: Organic waste from food processing, paper manufacturing, and other industries can be used as biomass.

Wastewater Sludge: Sludge from wastewater treatment plants can be processed to generate biogas through anaerobic digestion.

Processes of Converting Biomass into Energy

Biomass can be converted into usable forms of energy through various processes, each suited to different types of biomass and energy needs.

1: Combustion

Direct Combustion: Burning biomass in boilers or furnaces to produce heat and power. This is the most straightforward method of biomass conversion, commonly used for heating and industrial processes.

Co-Firing: Biomass is co-fired with coal in existing power plants to reduce greenhouse gas emissions and decrease reliance on fossil fuels.

2: Thermochemical Conversion

Pyrolysis: Heating biomass in the absence of oxygen to produce bio-oil, syngas, and biochar. Bio-oil can be further refined into biofuels, syngas can be used for power generation, and biochar can be used as a soil amendment.

Gasification: Partial oxidation of biomass at high temperatures to produce syngas, which can be used for electricity generation or as a feedstock for producing chemicals and fuels.

Torrefaction: Mild pyrolysis that produces a charcoal-like product, torrefied biomass, which is more energy-dense and easier to transport and store than raw biomass.

3: Biochemical Conversion

Anaerobic Digestion: Microorganisms break down organic matter in the absence of oxygen, producing biogas (a mixture of methane and carbon dioxide) and digestate. Biogas can be used for heating, electricity generation, or upgraded to biomethane for use as a vehicle fuel or injection into natural gas grids.

Fermentation: Microorganisms convert sugars and starches in biomass into ethanol, a liquid biofuel used for transportation. Lignocellulosic biomass, such as crop residues and wood, can also be converted to ethanol through advanced processes involving pretreatment and enzymatic hydrolysis.

4: Chemical Conversion

Transesterification: A chemical process that converts fats and oils into biodiesel, a renewable alternative to diesel fuel. Biodiesel can be produced from vegetable oils, animal fats, and waste cooking oil.

Sustainability of Biomass Energy

While biomass is a renewable energy source, its sustainability depends on several factors, including resource management, environmental impacts, and social considerations.

1: Resource Management

Sustainable Harvesting: Ensuring that biomass resources are harvested sustainably is crucial to maintaining their renewability. This involves practices like replanting trees, managing crop residues, and preventing deforestation.

Land Use: Competing demands for land can impact the sustainability of biomass. Energy crops should be grown on marginal or degraded lands to avoid displacing food production or natural ecosystems.

2: Environmental Impacts

Greenhouse Gas Emissions: While biomass combustion releases carbon dioxide, it is considered carbon-neutral if the carbon released is offset by the carbon absorbed during the growth of the biomass. However, emissions from land-use changes, transportation, and processing must also be considered.

Air Quality: Biomass combustion can produce pollutants such as particulate matter, nitrogen oxides, and volatile organic compounds. Advanced combustion technologies and emission controls are essential to minimize these impacts.

Water Use and Quality: Biomass production can affect water resources through irrigation, runoff, and water pollution. Sustainable farming and forestry practices can help mitigate these impacts.

3: Social Considerations

Energy Access and Security: Biomass energy can improve energy access and security, particularly in rural areas where biomass resources are abundant. It can provide decentralized energy solutions and create local jobs.

Food vs. Fuel Debate: The use of food crops for biofuel production can raise ethical and economic concerns, particularly in regions facing food insecurity. Prioritizing non-food biomass and using agricultural residues can help address this issue.

Future Potential and Innovations in Biomass Energy

Advancements in technology and innovative approaches are enhancing the potential of biomass energy to contribute to a sustainable energy future.

1: Advanced Biofuels

Cellulosic Ethanol: Produced from non-food biomass, such as agricultural residues and dedicated energy crops, cellulosic ethanol offers a more sustainable alternative to traditional biofuels.

Algae Biofuels: Algae can be cultivated to produce biofuels, offering high yields and the ability to grow on non-arable land using wastewater or saline water.

2: Integrated Biomass Systems

Biorefineries: Similar to petroleum refineries, biorefineries convert biomass into a range of products, including biofuels, biochemicals, and biomaterials. This integrated approach maximizes the value of biomass and reduces waste.

Combined Heat and Power (CHP): CHP systems generate both electricity and heat from biomass, improving overall energy efficiency and reducing greenhouse gas emissions.

3: Innovative Feedstocks

Waste-to-Energy: Utilizing organic waste, such as food waste and wastewater sludge, for energy production can address waste management challenges and provide a renewable energy source.

Second-Generation Feedstocks: Research is ongoing to develop new biomass feedstocks, such as perennial grasses and fast-growing trees, that require less water, fertilizer, and land.

Conclusion

Biomass energy, with its diverse sources and conversion processes, offers a versatile and renewable solution to the global energy challenge. By utilizing agricultural residues, forestry products, and organic waste, biomass can provide a sustainable alternative to fossil fuels for electricity, heat, and biofuels. However, ensuring the sustainability of biomass energy requires careful resource management, minimizing environmental impacts, and addressing social considerations. With advancements in technology and innovative approaches, biomass energy has the potential to play a significant role in a sustainable energy future, contributing to climate change mitigation, energy security, and rural development. By harnessing the power of organic materials, we can move towards a cleaner, more resilient energy system.

Chapter 9
Energy Storage and Grid Integration

Introduction to Energy Storage and Grid Integration

The integration of renewable energy sources into the existing power grid poses significant challenges and requires innovative solutions. Unlike traditional fossil fuel power plants, renewable energy sources such as solar and wind are intermittent and variable, creating the need for effective energy storage and advanced grid management. This chapter delves into the importance of energy storage technologies, such as batteries and pumped hydro, and explores the role of smart grids in ensuring the seamless integration of renewable energy into the power system.

Challenges of Integrating Renewable Energy into the Grid

1: Intermittency and Variability

Solar and Wind Power: Solar energy is dependent on sunlight, which varies throughout the day and is absent at night. Wind power is subject to fluctuating wind speeds and can be unpredictable.

Grid Stability: The variability of renewable energy can cause imbalances between supply and demand, leading to grid instability and potential outages.

2: Energy Storage Requirements

Peak Demand: Renewable energy generation may not always coincide with periods of peak electricity demand, necessitating the storage of excess energy for later use.

Grid Reliability: To maintain a reliable power supply, stored energy can be used to smooth out fluctuations in renewable energy output.

3: Infrastructure and Investment

Grid Modernization: Integrating renewable energy requires upgrades to the existing grid infrastructure, including transmission lines, substations, and control systems.

Investment Costs: The initial costs of developing and deploying energy storage systems and smart grid technologies can be high, requiring significant investment.

Energy Storage Technologies

Effective energy storage is crucial for mitigating the intermittency of renewable energy sources and ensuring a stable and reliable power supply.

1: Batteries

Lithium-Ion Batteries: Widely used in consumer electronics and electric vehicles, lithium-ion batteries are increasingly being deployed for grid-scale energy storage due to their high energy density and efficiency.

Flow Batteries: Flow batteries, such as vanadium redox flow batteries, store energy in liquid electrolytes and offer long cycle life and scalability for large-scale energy storage.

Solid-State Batteries: Emerging as a next-generation technology, solid-state batteries promise higher energy density, improved safety, and longer lifespans compared to conventional lithium-ion batteries.

2: Pumped Hydro Storage

How It Works: Pumped hydro storage involves pumping water from a lower reservoir to an upper reservoir during periods of excess electricity generation. When electricity demand is high, the water is released to flow back down through turbines, generating electricity.

Advantages: Pumped hydro storage is the most mature and widely used large-scale energy storage technology, offering high efficiency and long operational life.

Challenges: The construction of pumped hydro facilities requires suitable geographic locations with significant elevation differences, and it can have environmental impacts on local ecosystems.

3: Compressed Air Energy Storage (CAES)

How It Works: CAES systems store energy by compressing air into underground caverns or storage tanks. When electricity is needed, the compressed air is released, heated, and expanded through turbines to generate power.

Advantages: CAES can provide large-scale energy storage with relatively low operational costs and long lifespans.

Challenges: The efficiency of CAES is lower compared to other storage technologies, and suitable geological formations are required for underground storage.

4: Thermal Energy Storage

Molten Salt Storage: Used primarily in concentrated solar power (CSP) plants, molten salt storage involves heating a mixture of salts to store thermal energy, which can be used to generate steam and produce electricity when needed.

Phase Change Materials (PCMs): PCMs store energy by changing their physical state (e.g., from solid to liquid) and can be used for heating and cooling applications.

Advantages: Thermal energy storage is well-suited for integrating with solar thermal power plants and can provide high efficiency and long-duration storage.

Challenges: Thermal storage systems require careful management of temperature and material properties to ensure efficiency and reliability.

5: Flywheels

How They Work: Flywheels store energy in the form of rotational kinetic energy by spinning a rotor at high speeds. The stored energy can be quickly released to generate electricity when needed.

Advantages: Flywheels offer rapid response times, high power density, and long cycle life, making them suitable for short-term energy storage and grid stabilization.

Challenges: Flywheels have limited energy storage capacity and are primarily used for frequency regulation and short-duration applications.

The Role of Smart Grids

Smart grids are essential for the effective integration of renewable energy and the efficient management of energy storage systems. They incorporate advanced technologies to enhance grid reliability, flexibility, and efficiency.

1: Advanced Metering Infrastructure (AMI)

Smart Meters: Smart meters provide real-time data on electricity consumption, enabling consumers and utilities to monitor and manage energy usage more effectively.

Two-Way Communication: AMI enables two-way communication between utilities and consumers, facilitating demand response programs and real-time grid management.

2: Demand Response and Management

Demand Response Programs: These programs incentivize consumers to reduce or shift their electricity usage during peak demand periods, helping to balance supply and demand and reduce the need for additional power generation.

Automated Demand Management: Smart grid technologies can automatically adjust energy usage in response to grid conditions, optimizing energy consumption and reducing strain on the grid.

3: Distributed Energy Resources (DERs)

Integration of DERs: Smart grids enable the integration of distributed energy resources, such as rooftop solar panels, small wind turbines, and home energy storage systems, into the grid.

Microgrids: Microgrids are localized grids that can operate independently or in conjunction with the main grid. They enhance grid resilience and provide backup power during outages.

4: Grid Monitoring and Control

Real-Time Monitoring: Advanced sensors and monitoring systems provide real-time data on grid conditions, enabling utilities to detect and respond to issues more quickly.

Automated Control Systems: Smart grids use automated control systems to optimize grid operations, balance load and generation, and manage energy storage and distribution.

5: Electric Vehicles (EVs)

Vehicle-to-Grid (V2G) Technology: V2G technology allows electric vehicles to act as mobile energy storage units, providing power back to the grid during peak demand periods and enhancing grid stability.

Charging Infrastructure: The development of widespread EV charging infrastructure is essential for supporting the integration of electric vehicles into the smart grid.

Benefits of Energy Storage and Smart Grids

The integration of energy storage and smart grid technologies offers numerous benefits, enhancing the reliability, efficiency, and sustainability of the power system.

1: Enhanced Grid Stability and Reliability

Smoothing Renewable Energy Output: Energy storage systems can smooth out fluctuations in renewable energy generation, providing a stable and reliable power supply.

Backup Power: Energy storage can provide backup power during grid outages, enhancing grid resilience and reliability.

2: Increased Renewable Energy Penetration

Maximizing Renewable Utilization: Energy storage enables the storage of excess renewable energy for use during periods of low generation, maximizing the utilization of renewable resources.

Reducing Curtailment: By storing excess energy, storage systems can reduce the need to curtail renewable energy generation, improving overall system efficiency.

3: Improved Energy Efficiency

Demand Response and Management: Smart grids optimize energy consumption through demand response programs and automated management, reducing overall energy use and improving efficiency.

Reduced Transmission Losses: Distributed energy resources and localized energy storage reduce the need for long-distance electricity transmission, minimizing transmission losses.

4: Economic Benefits

Cost Savings: Energy storage and smart grid technologies can reduce the need for expensive peaking power plants and grid infrastructure upgrades, resulting in cost savings for utilities and consumers.

Job Creation: The development and deployment of energy storage and smart grid technologies create jobs in manufacturing, installation, maintenance, and operation.

5: Environmental Benefits

Reduced Greenhouse Gas Emissions: By enabling higher penetration of renewable energy and improving energy efficiency, energy storage and smart grids contribute to significant reductions in greenhouse gas emissions.

Enhanced Integration of Electric Vehicles: The integration of EVs into the smart grid supports the transition to cleaner transportation, further reducing emissions and improving air quality.

Conclusion

The integration of renewable energy into the existing power grid presents both challenges and opportunities. Energy storage technologies, such as batteries, pumped hydro, and thermal storage, are essential for managing the intermittency of renewable energy sources and ensuring a stable and reliable power supply. Smart grids, with their advanced monitoring, control, and communication capabilities, play a crucial role in optimizing grid operations, integrating distributed energy resources, and enhancing overall system efficiency. By investing in energy storage and smart grid technologies, we can accelerate the transition to a sustainable energy future, reducing our reliance on fossil fuels, lowering greenhouse gas emissions, and creating a more resilient and efficient power system.

Chapter 10
Policy and Legislation

Introduction to Policy and Legislation in Renewable Energy

The adoption of renewable energy is not solely driven by technological advancements and market forces; it is also significantly influenced by policies and legislation. Governments around the world have implemented a range of measures to promote the development and deployment of renewable energy technologies. This chapter explores the various international agreements, national policies, and incentives designed to accelerate the transition to renewable energy.

International Agreements

International agreements play a crucial role in setting the global agenda for renewable energy and climate change mitigation. These agreements provide a framework for cooperation and commitment among nations.

1: The Paris Agreement (2015)

Overview: The Paris Agreement, adopted under the United Nations Framework Convention on Climate Change (UNFCCC), aims to limit global warming to well below 2 degrees Celsius above pre-industrial levels, with efforts to limit the increase to 1.5 degrees Celsius.

Commitments: Countries are required to submit nationally determined contributions (NDCs) outlining their plans to reduce greenhouse gas emissions and increase renewable energy deployment.

Mechanisms: The agreement includes mechanisms for financial support, technology transfer, and capacity building to assist developing countries in achieving their renewable energy goals.

2: Sustainable Development Goals (SDGs)

Goal 7: Affordable and Clean Energy: The United Nations' SDGs include a specific goal to ensure access to affordable, reliable, sustainable, and modern energy for all. This goal emphasizes the importance of renewable energy in achieving sustainable development.

Target 7.2: This target aims to increase the share of renewable energy in the global energy mix substantially by 2030.

Target 7.A: This target focuses on enhancing international cooperation to facilitate access to clean energy research and technology, including renewable energy, energy efficiency, and advanced and cleaner fossil-fuel technology.

3: International Renewable Energy Agency (IRENA)

Mission: IRENA is an intergovernmental organization that supports countries in their transition to a sustainable energy future, promoting the widespread adoption and sustainable use of renewable energy.

Activities: IRENA provides policy advice, capacity building, and data and analysis to help countries develop effective renewable energy policies and strategies.

National Policies and Legislation

National policies and legislation are critical for creating a conducive environment for renewable energy development. These measures can include mandates, incentives, and regulations to encourage investment and deployment.

1: Renewable Portfolio Standards (RPS)

Definition: RPS are regulatory mandates that require a certain percentage of electricity to be generated from renewable sources.

Examples: In the United States, many states have implemented RPS, with varying targets and timelines. For instance, California's RPS requires 60% of electricity to come from renewable sources by 2030.

Impact: RPS have been instrumental in driving renewable energy deployment by creating a guaranteed market for renewable energy.

2: Feed-in Tariffs (FiTs)

Definition: FiTs are policies that guarantee a fixed, premium price for renewable energy producers for the electricity they generate and feed into the grid.

Examples: Germany's Renewable Energy Sources Act (EEG) introduced one of the most successful FiT programs, leading to a significant increase in solar and wind energy installations.

Impact: FiTs provide financial stability and long-term revenue certainty for renewable energy projects, encouraging investment and development.

3: Tax Incentives and Subsidies

Investment Tax Credits (ITCs): These credits allow renewable energy project developers to deduct a percentage of their investment costs from their taxes. For example, the U.S. offers an ITC for solar energy systems.

Production Tax Credits (PTCs): PTCs provide a per-kilowatt-hour tax credit for electricity generated from renewable energy sources. The U.S. PTC has been particularly influential in the growth of the wind energy sector.

Grants and Loans: Governments may offer grants and low-interest loans to support renewable energy projects. The European Union, through its Horizon 2020 program, provides funding for renewable energy research and development.

4: Net Metering

Definition: Net metering allows consumers who generate their own renewable energy (e.g., through rooftop solar panels) to feed excess electricity back into the grid and receive credit on their utility bills.

Examples: Many U.S. states have implemented net metering policies, encouraging residential and commercial adoption of solar energy.

Impact: Net metering makes renewable energy systems more economically attractive to consumers, driving higher adoption rates.

5: Renewable Energy Certificates (RECs)

Definition: RECs represent the environmental attributes of renewable energy generation and can be traded separately from the physical electricity.

Examples: The U.S. and several other countries have established REC markets to incentivize renewable energy production.

Impact: RECs provide an additional revenue stream for renewable energy producers and allow businesses and individuals to support renewable energy development.

Incentives and Financial Mechanisms

To further encourage the deployment of renewable energy, governments and financial institutions have developed various incentives and financial mechanisms.

1: Green Bonds

Definition: Green bonds are fixed-income securities issued to finance projects with environmental benefits, including renewable energy projects.

Examples: The World Bank and other financial institutions have issued green bonds to fund renewable energy projects worldwide.

Impact: Green bonds attract investment from institutional and retail investors, providing significant capital for renewable energy development.

2: Carbon Pricing

Carbon Taxes: Governments impose taxes on carbon emissions, creating an economic incentive to reduce emissions and invest in renewable energy. Countries like Sweden and Canada have implemented carbon taxes.

Emissions Trading Systems (ETS): ETS, such as the European Union Emissions Trading System, cap the total level of greenhouse gas emissions and allow industries to buy and sell emission allowances. Revenues generated from these systems can be used to support renewable energy projects.

Impact: Carbon pricing mechanisms internalize the environmental costs of fossil fuels, making renewable energy more competitive and attractive.

3: Public-Private Partnerships (PPPs)

Definition: PPPs involve collaboration between government agencies and private sector companies to finance and implement renewable energy projects.

Examples: The UK's Contracts for Difference (CfD) scheme is a PPP that provides long-term contracts to renewable energy developers, guaranteeing a fixed price for their electricity.

Impact: PPPs leverage public and private resources, reducing financial risks and accelerating renewable energy deployment.

Regional and Local Initiatives

In addition to national policies, regional and local governments play a vital role in promoting renewable energy through tailored initiatives and programs.

1: Community Energy Projects

Definition: Community energy projects involve local communities in the development and ownership of renewable energy installations, such as solar farms or wind turbines.

Examples: In Denmark, many wind energy projects are community-owned, fostering local support and investment.

Impact: Community energy projects enhance local engagement, generate economic benefits for communities, and increase renewable energy capacity.

2: Municipal Renewable Energy Programs

Definition: Municipal governments can implement programs to promote renewable energy within their jurisdictions, including incentives for rooftop solar installations, energy efficiency upgrades, and electric vehicle infrastructure.

Examples: Cities like San Francisco and New York have ambitious renewable energy goals and programs to support local renewable energy development.

Impact: Municipal programs contribute to national renewable energy targets, improve local air quality, and create green jobs.

3: Regional Collaborations

Definition: Regional collaborations involve multiple local governments working together to develop and implement renewable energy projects and policies.

Examples: The European Union's Clean Energy for All Europeans package encourages regional cooperation to achieve collective renewable energy targets.

Impact: Regional collaborations enable resource sharing, reduce costs, and enhance the effectiveness of renewable energy initiatives.

Challenges and Opportunities

While policies and legislation have been instrumental in advancing renewable energy, several challenges and opportunities remain.

1: Policy Consistency and Stability

Challenge: Inconsistent or rapidly changing policies can create uncertainty for investors and developers, hindering renewable energy deployment.

Opportunity: Long-term, stable policies provide a predictable environment for investment and encourage sustained growth in the renewable energy sector.

2: Access to Finance

Challenge: Securing financing for renewable energy projects can be challenging, particularly in developing countries and for small-scale projects.

Opportunity: Innovative financial mechanisms, such as green bonds, microfinance, and crowdfunding, can expand access to capital and support diverse renewable energy initiatives.

3: Equitable Transition

Challenge: The transition to renewable energy must be equitable, ensuring that all communities benefit and that displaced workers in the fossil fuel industry are supported.

Opportunity: Policies that promote job training, social inclusion, and community ownership can create a just transition and maximize the social benefits of renewable energy.

4: Technology and Innovation

Challenge: Continuous technological advancements are necessary to improve the efficiency and cost-effectiveness of renewable energy systems.

Opportunity: Policies that support research and development, innovation, and technology transfer can accelerate the deployment of cutting-edge renewable energy technologies.

Conclusion

Effective policies and legislation are crucial for driving the global transition to renewable energy. International agreements, national mandates, financial incentives, and regional initiatives create a supportive environment for renewable energy development. By addressing the challenges and leveraging the opportunities, governments can accelerate the adoption of renewable energy, reduce greenhouse gas emissions, and achieve a sustainable and resilient energy future.

Chapter 11
Economic Impacts

Introduction to Economic Impacts of Renewable Energy

The transition to renewable energy has profound economic implications, encompassing job creation, investment opportunities, and a comprehensive cost-benefit analysis. Understanding these economic aspects is essential for policymakers, businesses, and society as a whole. This chapter delves into the economic impacts of renewable energy, highlighting its potential to drive sustainable growth, enhance energy security, and foster a cleaner environment.

Job Creation in the Renewable Energy Sector

One of the most significant economic benefits of renewable energy is job creation. The renewable energy sector offers diverse employment opportunities across various segments, including manufacturing, installation, maintenance, and research and development.

1: Employment Growth

Global Employment Trends: According to the International Renewable Energy Agency (IRENA), the renewable energy sector employed approximately 11.5 million people worldwide in 2019, with projections indicating continued growth.

Regional Distribution: Countries like China, the United States, India, and members of the European Union are leading in renewable energy employment, driven by substantial investments in solar, wind, and bioenergy projects.

2: Types of Jobs

Manufacturing: The production of solar panels, wind turbines, and other renewable energy equipment requires skilled labor in manufacturing facilities.

Installation and Maintenance: Installation and maintenance of renewable energy systems create jobs for technicians, engineers, and electricians. For example, rooftop solar installations require local installers and maintenance personnel.

Research and Development (R&D): Innovation in renewable energy technologies generates employment for scientists, engineers, and researchers focused on improving efficiency, reducing costs, and developing new solutions.

3: Case Studies

Solar Energy: In the United States, the solar industry employed over 240,000 workers in 2019, with significant contributions from installation, manufacturing, and sales.

Wind Energy: The wind energy sector supports jobs in turbine manufacturing, project development, and operation and maintenance. In Denmark, a leading wind energy producer, the industry employs tens of thousands of people.

Investment Opportunities in Renewable Energy

The renewable energy sector presents lucrative investment opportunities for both public and private entities. These investments are crucial for scaling up renewable energy capacity and fostering economic growth.

1: Capital Investment

Global Investment Trends: Investment in renewable energy has surged in recent years. In 2019, global investment in renewable energy reached approximately $282 billion, with significant contributions from solar and wind energy projects.

Leading Investors: Major investors in renewable energy include governments, institutional investors, venture capital firms, and corporations committed to sustainability.

2: Green Bonds and Financial Instruments

Green Bonds: Green bonds are debt securities issued to finance environmentally friendly projects, including renewable energy. They have gained popularity as a means of raising capital for sustainable investments.

Impact Investment Funds: Impact investment funds focus on generating positive social and environmental outcomes alongside financial returns. These funds often prioritize renewable energy projects.

3: Public-Private Partnerships (PPPs)

Collaborative Models: PPPs enable collaboration between governments and private sector companies to finance and develop renewable energy projects. These partnerships leverage public funding and private expertise to achieve mutual goals.

Examples: The United Kingdom's Contracts for Difference (CfD) scheme is a PPP model that provides long-term contracts to renewable energy developers, ensuring price stability and encouraging investment.

Cost-Benefit Analysis of Transitioning to Renewable Energy

Transitioning from fossil fuels to renewable energy involves a detailed cost-benefit analysis to assess the economic, environmental, and social impacts.

1: Costs of Renewable Energy

Initial Investment: The upfront costs of renewable energy projects can be high, including expenses for technology, infrastructure, and grid integration.

Operational and Maintenance Costs: Renewable energy systems require regular maintenance to ensure optimal performance. While these costs are generally lower than those for fossil fuel plants, they must be considered in the overall analysis.

2: Economic Benefits

Lower Operating Costs: Renewable energy sources, such as solar and wind, have minimal fuel costs compared to fossil fuels, leading to lower operating expenses over time.

Energy Independence: Investing in domestic renewable energy reduces dependence on imported fossil fuels, enhancing energy security and stabilizing energy prices.

Economic Diversification: Renewable energy development promotes economic diversification by creating new industries and reducing reliance on volatile fossil fuel markets.

3: Environmental and Social Benefits

Reduced Greenhouse Gas Emissions: Transitioning to renewable energy significantly lowers greenhouse gas emissions, mitigating climate change and its associated economic costs.

Improved Public Health: Reducing air pollution from fossil fuel combustion improves public health, leading to lower healthcare costs and increased productivity.

Social Equity: Renewable energy projects can bring economic development to underserved and rural communities, providing access to clean energy and creating local jobs.

4: Comparative Analysis

Renewable vs. Fossil Fuels: A comprehensive analysis comparing the long-term costs and benefits of renewable energy and fossil fuels shows that renewable energy offers substantial advantages in terms of sustainability, economic stability, and environmental impact.

Levelized Cost of Energy (LCOE): LCOE is a metric used to compare the total costs of different energy sources over their lifetimes. Recent trends show that the LCOE for solar and wind energy has decreased significantly, making them competitive with or cheaper than fossil fuels.

Economic Resilience and Renewable Energy

Renewable energy contributes to economic resilience by reducing vulnerability to energy price fluctuations and enhancing energy security.

1: Energy Price Stability

Renewable Energy Pricing: Unlike fossil fuels, renewable energy sources have stable and predictable pricing, as they are not subject to market volatility and geopolitical risks.

Impact on Consumers: Stable energy prices benefit consumers by reducing electricity costs and shielding them from price spikes associated with fossil fuels.

2: Energy Security

Domestic Production: Renewable energy projects often rely on local resources, such as sunlight, wind, and biomass, reducing dependence on imported fuels and enhancing national energy security.

Supply Chain Resilience: Developing renewable energy infrastructure strengthens supply chain resilience by diversifying energy sources and reducing exposure to global supply disruptions.

3: Economic Diversification and Innovation

New Markets: The growth of the renewable energy sector creates new markets and business opportunities, fostering economic diversification and innovation.

Technological Advancements: Investment in renewable energy drives technological advancements and research, leading to the development of more efficient and cost-effective solutions.

Challenges and Opportunities

While the economic impacts of renewable energy are largely positive, several challenges and opportunities must be addressed to maximize its potential.

1: Challenges

Financing and Investment: Securing adequate financing for large-scale renewable energy projects can be challenging, particularly in developing countries with limited access to capital.

Grid Integration: Integrating renewable energy into existing power grids requires significant infrastructure upgrades and investments in energy storage and grid management technologies.

Policy and Regulatory Barriers: Inconsistent policies, regulatory hurdles, and lack of supportive legislation can impede renewable energy development and investment.

2: Opportunities

Technological Innovation: Continued investment in research and development can drive technological innovation, reducing costs and improving the efficiency of renewable energy systems.

Public Awareness and Support: Increasing public awareness and support for renewable energy can drive policy changes and encourage investment from individuals, businesses, and governments.

International Collaboration: Collaborative efforts between countries can accelerate the global transition to renewable energy, leveraging shared knowledge, resources, and technology.

Conclusion

The economic impacts of renewable energy are multifaceted, encompassing job creation, investment opportunities, and a detailed cost-benefit analysis. Transitioning to renewable energy offers significant economic, environmental, and social benefits, contributing to sustainable growth, energy security, and climate change mitigation. By addressing challenges and leveraging opportunities, governments, businesses, and communities can harness the full potential of renewable energy, creating a resilient and prosperous future.

Chapter 12
Environmental and Social Benefits

Introduction to Environmental and Social Benefits of Renewable Energy

Renewable energy sources offer a multitude of environmental and social benefits that extend far beyond their economic advantages. This chapter explores the profound positive impacts renewable energy can have on the environment and society, focusing on reduced air pollution, improved public health, and enhanced energy security. By understanding these benefits, we can better appreciate the comprehensive value of transitioning to renewable energy.

Reduced Air Pollution

One of the most significant environmental benefits of renewable energy is the reduction of air pollution. Fossil fuel combustion is a major source of air pollutants, including sulfur dioxide (SO_2), nitrogen oxides (NO_x), particulate matter (PM), and volatile organic compounds (VOCs). Renewable energy sources, such as solar, wind, and hydroelectric power, do not produce these harmful emissions.

1: Reduction of Greenhouse Gas Emissions

Carbon Dioxide (CO_2): Renewable energy significantly reduces CO_2 emissions, a primary greenhouse gas contributing to climate change. For example, solar and wind energy produce no CO_2 during operation.

Methane (CH_4): Bioenergy projects, when managed sustainably, can reduce methane emissions from agricultural waste and landfills.

2: Decrease in Harmful Pollutants

Sulfur Dioxide (SO_2): SO_2 emissions, primarily from coal combustion, cause acid rain and respiratory problems. Switching to renewable energy eliminates these emissions.

Nitrogen Oxides (NO_x): NO_x contribute to smog formation and respiratory issues. Renewable energy sources, particularly wind and solar, do not emit NO_x.

Particulate Matter (PM): Fossil fuel burning releases fine particulate matter, which can penetrate deep into the lungs and cause serious health problems. Renewable energy sources do not generate PM during operation.

3: Impact on Ecosystems

Water Quality: Reduced air pollution from renewable energy projects leads to fewer acid rain events, benefiting aquatic ecosystems and water quality.

Soil Health: Decreased air pollution also reduces the deposition of harmful substances onto soil, improving soil health and agricultural productivity.

Improved Public Health

The reduction of air pollution from renewable energy sources has a direct and significant impact on public health. Cleaner air leads to fewer respiratory and cardiovascular diseases, improving overall health outcomes and reducing healthcare costs.

1: Respiratory Health

Asthma and Lung Disease: Lower levels of air pollutants, such as NO_x and PM, result in fewer cases of asthma, chronic bronchitis, and other respiratory conditions.

Children's Health: Children are particularly vulnerable to air pollution. Renewable energy reduces exposure to harmful pollutants, leading to healthier childhood development.

2: Cardiovascular Health

Heart Disease and Stroke: Air pollution is linked to increased risks of heart disease and stroke. By reducing emissions, renewable energy helps lower these health risks.

Blood Pressure: Cleaner air can also contribute to lower blood pressure, reducing the incidence of hypertension and related conditions.

3: Economic Benefits of Improved Health

Healthcare Savings: Fewer health problems due to air pollution lead to reduced healthcare costs, benefiting both individuals and public health systems.

Increased Productivity: Healthier populations are more productive, with fewer sick days and higher overall economic output.

Enhanced Energy Security

Renewable energy sources contribute to enhanced energy security by diversifying energy supplies and reducing dependence on imported fuels. This leads to greater stability and resilience in the energy system.

1: Energy Independence

Domestic Energy Production: Renewable energy projects often rely on local resources, such as sunlight, wind, and water. This reduces dependence on imported fossil fuels and enhances national energy security.

Reduction of Geopolitical Risks: By decreasing reliance on fossil fuel imports, countries can mitigate the risks associated with geopolitical tensions and energy supply disruptions.

2: Resilience to Price Volatility

Stable Energy Costs: Renewable energy sources have stable and predictable costs, unlike fossil fuels, which are subject to price fluctuations. This stability benefits consumers and businesses by providing predictable energy costs.

Mitigation of Economic Shocks: Renewable energy can help insulate economies from the economic shocks caused by sudden increases in fossil fuel prices, enhancing overall economic resilience.

3: Community Empowerment

Local Energy Production: Community-based renewable energy projects empower local communities by giving them control over their energy supply. This can lead to increased energy access and economic development in rural and underserved areas.

Energy Cooperatives: Energy cooperatives and community-owned renewable energy projects promote local investment and job creation, strengthening community ties and fostering economic self-sufficiency.

Environmental Preservation

Renewable energy also contributes to environmental preservation by reducing the environmental footprint of energy production.

1: Conservation of Natural Resources

Reduced Water Usage: Unlike fossil fuel and nuclear power plants, which require significant amounts of water for cooling, renewable energy technologies like solar and wind use minimal water. This conserves water resources and reduces the environmental impact on aquatic ecosystems.

Land Use Efficiency: While large-scale renewable energy projects do require land, they often coexist with agricultural activities or use previously disturbed lands, minimizing their impact on natural habitats.

2: Biodiversity Protection

Habitat Preservation: By reducing air and water pollution, renewable energy helps protect habitats and biodiversity. This is particularly important for sensitive ecosystems and endangered species.

Sustainable Bioenergy: When managed sustainably, bioenergy can provide a renewable energy source while also promoting biodiversity through practices such as agroforestry and sustainable forestry management.

Social Equity and Inclusion

The transition to renewable energy offers opportunities for social equity and inclusion, particularly for marginalized and underserved communities.

1: Energy Access

Rural Electrification: Renewable energy technologies, such as off-grid solar systems and mini-grids, can provide electricity to remote and rural areas that lack access to traditional power grids. This improves quality of life and economic opportunities for these communities.

Affordable Energy: As renewable energy costs continue to decline, it becomes a more affordable option for low-income households, reducing energy poverty and enhancing social equity.

2: Economic Opportunities

Job Creation: Renewable energy projects create jobs in local communities, providing employment opportunities and economic benefits. This is especially important in regions transitioning away from fossil fuel industries.

Community Ownership: Community-owned renewable energy projects can generate income and economic benefits for local residents, promoting inclusive economic development.

3: Climate Justice

Vulnerable Communities: Marginalized communities are often disproportionately affected by the impacts of climate change and environmental degradation. Renewable energy can help mitigate these impacts by reducing greenhouse gas emissions and promoting sustainable development.

Equitable Transition: Ensuring that the transition to renewable energy is equitable involves providing support and opportunities for workers and communities impacted by the decline of fossil fuel industries. This includes job training, social safety nets, and investment in economic diversification.

Challenges and Opportunities

While the environmental and social benefits of renewable energy are significant, there are challenges and opportunities that must be addressed to fully realize these benefits.

1: Challenges

Initial Costs: The upfront costs of renewable energy projects can be a barrier, particularly for low-income communities and developing countries. Access to financing and supportive policies are essential to overcoming this challenge.

Infrastructure and Grid Integration: Integrating renewable energy into existing power grids requires significant infrastructure upgrades and investments in energy storage and grid management technologies.

Land and Resource Conflicts: Large-scale renewable energy projects can sometimes lead to land and resource conflicts, particularly in areas with competing land uses or limited natural resources. It is important to address these conflicts through inclusive planning and stakeholder engagement.

2: Opportunities

Technological Innovation: Continued investment in research and development can drive technological innovation, reducing costs and improving the efficiency of renewable energy systems.

Public Awareness and Support: Increasing public awareness and support for renewable energy can drive policy changes and encourage investment from individuals, businesses, and governments.

International Collaboration: Collaborative efforts between countries can accelerate the global transition to renewable energy, leveraging shared knowledge, resources, and technology.

Conclusion

The environmental and social benefits of renewable energy are far-reaching and transformative. By reducing air pollution, improving public health, enhancing energy security, preserving natural resources, and promoting social equity, renewable energy offers a comprehensive solution to many of the world's most pressing challenges. Embracing renewable energy is not only a crucial step toward mitigating climate change but also a pathway to a more sustainable, equitable, and prosperous future for all.

Chapter 13
Case Studies
Successful Renewable Energy Projects

Introduction to Successful Renewable Energy Projects

Around the world, numerous renewable energy projects have demonstrated significant environmental, economic, and social benefits. This chapter examines some of these successful projects, highlighting their implementation, impacts, and lessons learned. Through real-world examples, we gain insights into how renewable energy can drive sustainable development and transform local communities.

Case Study 1: The Alta Wind Energy Center, USA

Project Overview

Location: Tehachapi, California, USA

Capacity: Approximately 1,550 megawatts (MW)

Technology: Wind turbines

Implementation

The Alta Wind Energy Center (AWEC) is one of the largest wind farms in the world. Located in the Tehachapi Pass, a region known for its high wind speeds, the project leverages the area's natural wind resources to generate electricity. The AWEC consists of multiple phases, each involving the installation of additional wind turbines to increase capacity.

Impacts

Environmental: The AWEC significantly reduces greenhouse gas emissions by displacing fossil fuel-based power generation. It produces enough electricity to power approximately 450,000 homes annually, avoiding millions of tons of CO_2 emissions.

Economic: The project has created numerous jobs in construction, operation, and maintenance. It has also generated economic benefits for the local community through lease payments to landowners and increased tax revenues.

Community: The AWEC has fostered community engagement and support through educational programs and partnerships with local schools and organizations.

Lessons Learned

Strategic Location: Selecting a location with optimal wind resources is crucial for maximizing energy output and project viability.

Community Involvement: Engaging with local communities and stakeholders early in the project can build support and address potential concerns.

Case Study 2: Noor Solar Complex, Morocco

Project Overview

Location: Ouarzazate, Morocco

Capacity: 580 MW (combined capacity of Noor I, II, and III)

Technology: Concentrated solar power (CSP) and photovoltaic (PV) panels

Implementation

The Noor Solar Complex is a flagship project in Morocco's renewable energy strategy. It combines CSP and PV technologies to harness the region's abundant solar resources. The complex includes the Noor I, II, and III plants, each employing different CSP technologies to maximize efficiency and output.

Impacts

Environmental: The Noor Solar Complex significantly reduces Morocco's reliance on fossil fuels, cutting CO_2 emissions by approximately 760,000 tons annually. It provides clean energy to around 1.1 million people.

Economic: The project has attracted substantial foreign investment and created numerous jobs in construction, operation, and maintenance. It has also spurred local economic development by promoting the growth of related industries.

Community: The project has engaged local communities through educational initiatives and employment opportunities. It has also contributed to infrastructure development, such as roads and water systems.

Lessons Learned

Technology Diversification: Combining different solar technologies can enhance overall efficiency and reliability.

Public-Private Partnerships: Effective collaboration between public and private entities can facilitate financing and implementation.

Case Study 3: Samsø Renewable Energy Island, Denmark

Project Overview

Location: Samsø Island, Denmark

Capacity: Approximately 28 MW (wind), 2.5 MW (solar), and 11 MW (biomass)

Technology: Wind turbines, solar panels, biomass plants

Implementation

Samsø Island embarked on an ambitious plan to become 100% renewable within ten years. The project involved installing wind turbines, solar panels, and biomass plants to meet the island's energy needs. Local residents and cooperatives played a significant role in financing and managing the installations.

Impacts

Environmental: Samsø Island now generates more renewable energy than it consumes, with surplus electricity exported to the mainland. The project has drastically reduced CO_2 emissions, contributing to Denmark's national climate goals.

Economic: The project has created jobs in renewable energy and related sectors, boosting the local economy. It has also reduced energy costs for residents and businesses.

Community: Samsø's transition to renewable energy has fostered a strong sense of community pride and ownership. The project has become a model for other regions seeking to achieve energy independence.

Lessons Learned

Community Ownership: Involving local residents in project ownership and decision-making can enhance support and ensure long-term success.

Integrated Approach: Combining multiple renewable energy sources can create a resilient and self-sufficient energy system.

Case Study 4: The Hornsdale Power Reserve, Australia

Project Overview

Location: Jamestown, South Australia

Capacity: 150 MW / 194 MWh

Technology: Lithium-ion battery storage

Implementation

The Hornsdale Power Reserve, known as the "Tesla Big Battery," is one of the world's largest lithium-ion battery installations. The project was implemented to provide grid stability, store excess renewable energy, and support peak demand periods.

Impacts

Environmental: The battery storage system enhances the integration of renewable energy into the grid, reducing reliance on fossil fuel-based peaker plants. It helps smooth out fluctuations in solar and wind power generation.

Economic: The project has reduced electricity prices and provided savings on grid services. It has also created jobs in construction, operation, and maintenance.

Community: The battery system has improved grid reliability and reduced the frequency and duration of power outages, benefiting local communities and businesses.

Lessons Learned

Energy Storage: Advanced energy storage technologies are essential for integrating high levels of renewable energy into the grid.

Rapid Implementation: Public-private partnerships and streamlined regulatory processes can facilitate the rapid deployment of large-scale renewable energy projects.

Case Study 5: Krafla Geothermal Power Station, Iceland

Project Overview

Location: Krafla, Iceland

Capacity: 60 MW

Technology: Geothermal power

Implementation

The Krafla Geothermal Power Station utilizes Iceland's abundant geothermal resources to generate electricity. The project involves drilling deep wells to access geothermal reservoirs, with steam and hot water used to drive turbines.

Impacts

Environmental: Geothermal power is a low-emission energy source. The Krafla station significantly reduces CO_2 emissions compared to fossil fuel-based power plants.

Economic: The project provides stable and affordable electricity, supporting Iceland's energy-intensive industries and promoting economic growth.

Community: The geothermal project has created jobs and contributed to local infrastructure development, including roads and utilities.

Lessons Learned

Resource Management: Sustainable management of geothermal resources is crucial to ensure long-term viability and minimize environmental impact.

Local Benefits: Geothermal projects can drive local economic development and improve infrastructure, benefiting surrounding communities.

Conclusion

These case studies illustrate the diverse ways renewable energy projects are being successfully implemented around the world. From large-scale wind farms and solar complexes to community-driven initiatives and advanced energy storage systems, renewable energy is transforming local communities and driving sustainable development. By learning from these examples, we can better understand the potential of renewable energy to address global challenges and create a cleaner, more resilient future.

Chapter 14
Technological Innovations

Introduction to Technological Innovations in Renewable Energy

Technological advancements play a critical role in enhancing the efficiency, reliability, and affordability of renewable energy sources. This chapter explores the latest innovations in solar panels, wind turbine designs, and energy storage solutions, highlighting how these developments are transforming the renewable energy landscape and accelerating the transition to a sustainable energy future.

Solar Panel Innovations

1. Perovskite Solar Cells

Overview: Perovskite solar cells (PSCs) are a new class of solar cells that use perovskite-structured compounds as the light-absorbing layer. They have shown remarkable efficiency improvements and have the potential to surpass traditional silicon-based solar cells.

Advantages:

High Efficiency: PSCs have achieved efficiencies above 25% in laboratory settings, with the potential for further improvements.

Low Cost: The materials and manufacturing processes for PSCs are less expensive than those for silicon solar cells.

Flexibility: PSCs can be produced on flexible substrates, allowing for a wide range of applications, including building-integrated photovoltaics (BIPV) and wearable devices.

Challenges:

Stability: PSCs currently have issues with long-term stability and degradation under environmental conditions.

Scalability: Scaling up production to commercial levels while maintaining high efficiency and stability is a key challenge.

2. Bifacial Solar Panels

Overview: Bifacial solar panels can capture sunlight on both sides of the panel, increasing their overall energy generation.

Advantages:

Increased Energy Yield: By capturing reflected and diffused light from the rear side, bifacial panels can generate up to 30% more energy than traditional monofacial panels.

Durability: Bifacial panels often have more robust designs to withstand environmental conditions.

Versatility: They can be used in various configurations, including vertical installations and over reflective surfaces like snow or water.

Challenges:

Installation Costs: The installation and mounting systems for bifacial panels can be more complex and costly.

Performance Variability: The performance of bifacial panels depends on the reflectivity of the ground surface and other environmental factors.

3. Thin-Film Solar Cells

Overview: Thin-film solar cells are made by depositing one or more thin layers of photovoltaic material onto a substrate. Common materials include cadmium telluride (CdTe) and copper indium gallium selenide (CIGS).

Advantages:

Lightweight and Flexible: Thin-film solar cells are lighter and more flexible than traditional silicon panels, making them suitable for a variety of applications, including portable and building-integrated systems.

Lower Material Use: They require less raw material, which can reduce costs and environmental impact.

Good Performance in Low Light: Thin-film solar cells often perform better in low-light conditions and at higher temperatures.

Challenges:

Lower Efficiency: While they have improved over time, thin-film solar cells generally have lower efficiencies compared to silicon-based cells.

Toxic Materials: Some thin-film technologies use toxic materials (e.g., cadmium), raising environmental and health concerns.

Wind Turbine Innovations

1. Offshore Wind Turbines

Overview: Offshore wind turbines are installed in bodies of water, where wind speeds are typically higher and more consistent than on land.

Advantages:

Higher Capacity Factors: Offshore wind farms can achieve higher capacity factors due to stronger and more consistent winds.

Reduced Land Use: Offshore installations do not compete with land use for agriculture or urban development.

Large-Scale Deployment: Offshore turbines can be larger than onshore turbines, increasing their energy output.

Challenges:

High Costs: Installation and maintenance of offshore wind turbines are more expensive due to the harsh marine environment.

Environmental Impact: Offshore wind projects can impact marine ecosystems and require careful planning to minimize disruption.

2. Floating Wind Turbines

Overview: Floating wind turbines are designed to be deployed in deep waters where traditional fixed-bottom turbines are not feasible.

Advantages:

Access to Deep Water Sites: Floating turbines can be installed in deeper waters with stronger winds, increasing energy potential.

Reduced Visual Impact: They can be placed further offshore, reducing visual impact and potential conflicts with other land uses.

Scalability: Floating turbines offer scalability and flexibility in site selection.

Challenges:

Technology Maturity: Floating wind technology is still in the early stages of commercial deployment and faces technical and financial challenges.

Anchoring and Stability: Ensuring the stability and durability of floating platforms in harsh ocean conditions is a key engineering challenge.

3. Vertical Axis Wind Turbines (VAWTs)

Overview: VAWTs have a vertical rotor shaft and can capture wind from any direction, unlike traditional horizontal axis wind turbines (HAWTs).

Advantages:

Omni-Directional: VAWTs can capture wind from all directions, eliminating the need for yaw mechanisms.

Lower Noise Levels: They tend to operate more quietly than HAWTs, making them suitable for urban and residential areas.

Ease of Maintenance: Key components are located closer to the ground, facilitating easier maintenance.

Challenges:

Lower Efficiency: VAWTs generally have lower efficiencies compared to HAWTs, limiting their widespread adoption.

Structural Challenges: They face issues with structural stress and fatigue, particularly in larger designs.

Energy Storage Solutions

1. Lithium-Ion Batteries

Overview: Lithium-ion batteries are the most widely used energy storage technology, known for their high energy density and efficiency.

Advantages:

High Energy Density: Lithium-ion batteries store more energy per unit of weight and volume compared to other battery technologies.

Long Cycle Life: They have a long cycle life, making them suitable for grid storage and electric vehicles.

Fast Response Time: Lithium-ion batteries can quickly respond to changes in energy demand, providing grid stability and supporting renewable integration.

Challenges:

Cost: Although costs have decreased, lithium-ion batteries are still relatively expensive.

Resource Constraints: The supply of lithium and other critical materials (e.g., cobalt) can be limited and subject to geopolitical risks.

Safety Concerns: There are safety concerns related to thermal runaway and fire risk.

2. Solid-State Batteries

Overview: Solid-state batteries use a solid electrolyte instead of a liquid one, offering potential improvements in energy density and safety.

Advantages:

Higher Energy Density: Solid-state batteries can achieve higher energy densities, extending the range of electric vehicles and providing more compact energy storage solutions.

Enhanced Safety: The solid electrolyte reduces the risk of leaks and fires, improving overall safety.

Longer Lifespan: Solid-state batteries are expected to have longer lifespans and better performance at high temperatures.

Challenges:

Manufacturing Complexity: Producing solid-state batteries at scale remains a significant technical challenge.

Cost: Current production methods are costly, and achieving cost parity with lithium-ion batteries is a key hurdle.

3. Flow Batteries

Overview: Flow batteries store energy in liquid electrolytes contained in external tanks, allowing for scalable and flexible energy storage solutions.

Advantages:

Scalability: The capacity of flow batteries can be easily scaled by increasing the size of the electrolyte tanks.

Long Cycle Life: Flow batteries have a long cycle life and can withstand many charge-discharge cycles without significant degradation.

Safety: They are generally safer than lithium-ion batteries, with a lower risk of thermal runaway.

Challenges:

Low Energy Density: Flow batteries have lower energy densities compared to lithium-ion batteries, making them less suitable for applications requiring high energy density.

Complexity and Cost: The complexity of the system and the cost of materials, such as vanadium in vanadium redox flow batteries, can be barriers to widespread adoption.

4. Hydrogen Storage

Overview: Hydrogen can be produced using renewable energy and stored for later use in fuel cells or combustion processes.

Advantages:

High Energy Content: Hydrogen has a high energy content per unit weight, making it suitable for long-term and large-scale energy storage.

Versatility: Hydrogen can be used in various applications, including transportation, industrial processes, and power generation.

Renewable Production: When produced through electrolysis using renewable energy (green hydrogen), hydrogen is a clean and sustainable energy carrier.

Challenges:

Efficiency: The overall efficiency of hydrogen production, storage, and conversion is lower than that of direct electricity storage.

Infrastructure: Developing the necessary infrastructure for hydrogen production, storage, and distribution requires significant investment.

Cost: The cost of green hydrogen production is currently higher than that of fossil fuel-based hydrogen (gray hydrogen).

Conclusion

Technological innovations in renewable energy are driving significant advancements in solar power, wind energy, and energy storage solutions. These innovations are not only improving the efficiency and cost-effectiveness of renewable energy systems but also expanding their applicability and scalability. By overcoming the challenges associated with these technologies and continuing to invest in research and development, we can accelerate the transition to a sustainable energy future. The ongoing evolution of renewable energy technologies promises to play a crucial role in mitigating climate change and securing a cleaner, more resilient energy system for future generations.

Chapter 15

Barriers to Adoption

Introduction to Barriers to Adoption

The transition to renewable energy is essential for mitigating climate change and ensuring a sustainable future. However, several barriers hinder the widespread adoption of renewable energy technologies. This chapter delves into the technological, economic, and political challenges that must be overcome to accelerate the global shift to renewable energy.

Technological Limitations

1. Intermittency and Reliability

Overview: Renewable energy sources like solar and wind are intermittent, producing electricity only when the sun is shining or the wind is blowing.

Challenges:

Grid Stability: Integrating large amounts of intermittent energy can destabilize the grid, requiring advanced management and balancing techniques.

Storage Solutions: Effective energy storage systems are needed to store excess energy and supply it during periods of low generation. Current storage technologies, such as batteries, are expensive and have limited capacity.

Predictability: Variability in weather conditions makes it challenging to predict energy generation, complicating grid management and planning.

2. Technological Maturity

Overview: Some renewable energy technologies are still in the early stages of development and have not yet reached commercial maturity.

Challenges:

Efficiency and Cost: Emerging technologies, such as advanced solar cells, floating wind turbines, and wave energy, need further development to improve efficiency and reduce costs.

Scaling Up: Scaling up production from pilot projects to commercial-scale operations involves significant technical and financial hurdles.

R&D Investment: Sustained investment in research and development is necessary to advance these technologies and make them viable alternatives to conventional energy sources.

3. Grid Infrastructure

Overview: The existing grid infrastructure is often outdated and not designed to accommodate the decentralized and variable nature of renewable energy.

Challenges:

Modernization: Upgrading the grid to handle distributed generation, smart grid technologies, and advanced metering is costly and time-consuming.

Transmission Capacity: Renewable energy sources are often located far from population centers, requiring extensive transmission networks to deliver electricity to end users.

Grid Integration: Ensuring seamless integration of renewable energy into the grid involves complex technical and regulatory challenges.

Economic Challenges

1. High Initial Costs

Overview: The upfront capital costs of renewable energy projects are often higher than those of conventional fossil fuel plants.

Challenges:

Financing: Securing financing for renewable energy projects can be difficult, particularly in developing countries or regions with limited access to capital.

Cost Competitiveness: Despite falling prices, renewable energy technologies must compete with established fossil fuel infrastructure, which can be cheaper in the short term.

Economic Incentives: Governments and financial institutions need to provide incentives, such as subsidies, tax credits, and low-interest loans, to make renewable energy projects more attractive to investors.

2. Market Dynamics

Overview: The energy market is often dominated by fossil fuel interests, creating economic barriers for renewable energy adoption.

Challenges:

Subsidies for Fossil Fuels: Fossil fuel industries receive significant subsidies, distorting the market and making it difficult for renewable energy to compete on a level playing field.

Price Volatility: The prices of fossil fuels can fluctuate widely, affecting the competitiveness of renewable energy.

Market Structures: Traditional energy market structures and regulations may not be conducive to the integration of renewable energy, requiring reforms to support a transition to a low-carbon economy.

3. Economic Disruption

Overview: The shift to renewable energy can disrupt existing economic structures and industries, leading to resistance from affected stakeholders.

Challenges:

Job Losses: Transitioning from fossil fuels to renewable energy can result in job losses in traditional energy sectors, necessitating retraining and support for displaced workers.

Economic Impact on Communities: Regions dependent on fossil fuel extraction and production may face economic decline and social challenges as industries contract.

Investment in Transition: Significant investment is needed to manage the transition, including developing new supply chains, infrastructure, and training programs for workers.

Political Resistance

1. Policy and Regulatory Barriers

Overview: Inconsistent or unfavorable policies and regulations can hinder the adoption of renewable energy.

Challenges:

Lack of Supportive Policies: Some governments lack the political will or awareness to implement policies that promote renewable energy, such as feed-in tariffs, renewable portfolio standards, and carbon pricing.

Regulatory Uncertainty: Frequent changes in policies and regulations create uncertainty for investors and developers, discouraging long-term investments in renewable energy.

Bureaucratic Hurdles: Complex permitting processes, land use regulations, and other bureaucratic obstacles can delay or block renewable energy projects.

2. Influence of Fossil Fuel Interests

Overview: Powerful fossil fuel industries and their political allies often resist efforts to transition to renewable energy.

Challenges:

Lobbying and Political Influence: Fossil fuel companies wield significant influence through lobbying and campaign contributions, shaping policies that favor their interests.

Public Perception: Misleading information campaigns by fossil fuel interests can sway public opinion against renewable energy and climate policies.

Economic Dependence: Governments and regions heavily dependent on fossil fuel revenues may resist transitioning to renewable energy to protect their economic interests.

3. Geopolitical Factors

Overview: International relations and geopolitical considerations can impact the adoption of renewable energy.

Challenges:

Energy Security: Countries reliant on fossil fuel imports may prioritize energy security over transitioning to renewable energy, seeking to maintain stable and affordable energy supplies.

International Competition: Geopolitical competition for technological leadership in renewable energy can affect cooperation and coordination on global climate initiatives.

Trade Policies: Trade barriers and protectionist policies can hinder the global exchange of renewable energy technologies and materials.

Conclusion

The widespread adoption of renewable energy faces numerous barriers, including technological limitations, economic challenges, and political resistance. Overcoming these obstacles requires a coordinated effort involving technological innovation, supportive policies, financial incentives, and public engagement. By addressing these challenges, we can accelerate the transition to a sustainable energy future and mitigate the impacts of climate change. The path to renewable energy adoption is complex, but with concerted efforts and strategic investments, it is achievable and essential for a resilient and sustainable world.

Chapter 16

The Role of Corporations and Industry

Introduction to Corporate Embrace of Renewable Energy

Corporations and industries play a pivotal role in the global transition to renewable energy. With increasing awareness of environmental sustainability and the economic benefits of renewable energy, many companies are integrating renewable energy into their operations and strategies. This chapter explores corporate sustainability initiatives, green certifications, and the role of renewable energy in corporate social responsibility (CSR).

Corporate Sustainability Initiatives

1. Corporate Renewable Energy Commitments

Overview: Many corporations have set ambitious renewable energy goals as part of their sustainability strategies.

Examples:

Google: Google has been a leader in corporate renewable energy, achieving its goal of matching 100% of its global energy consumption with renewable energy purchases.

Apple: Apple has committed to powering all its facilities worldwide with 100% renewable energy and has achieved this goal for its operations in 43 countries.

IKEA: IKEA aims to become climate positive by 2030, which includes generating more renewable energy than it consumes.

2. Corporate Power Purchase Agreements (PPAs)

Overview: Power Purchase Agreements (PPAs) are long-term contracts between corporations and renewable energy developers to purchase electricity directly from renewable energy projects.

Advantages:

Stable Energy Costs: PPAs provide long-term price stability for energy costs, protecting companies from market volatility.

Greenhouse Gas Reduction: By sourcing renewable energy through PPAs, corporations can significantly reduce their carbon footprints.

Market Signal: Large PPAs signal strong demand for renewable energy, encouraging further development and investment in the sector.

Challenges:

Contract Complexity: Negotiating and managing PPAs can be complex and resource-intensive.

Regulatory Barriers: Regulatory frameworks in some regions may not support the use of PPAs.

3. Onsite Renewable Energy Generation

Overview: Some corporations invest in onsite renewable energy systems, such as solar panels and wind turbines, to generate their own electricity.

Examples:

Walmart: Walmart has installed solar panels on the rooftops of many of its stores and distribution centers, aiming to be powered by 100% renewable energy.

Google: In addition to PPAs, Google has invested in onsite renewable energy projects, including solar and wind installations at its data centers.

General Motors: General Motors has installed solar arrays and wind turbines at several of its manufacturing facilities.

Green Certifications and Standards

1. LEED Certification

Overview: Leadership in Energy and Environmental Design (LEED) is a widely recognized green building certification that promotes sustainable building practices.

Benefits:

Energy Efficiency: LEED-certified buildings are designed to be energy-efficient, often incorporating renewable energy systems.

Market Value: LEED certification can enhance a company's brand and increase the market value of its properties.

Environmental Impact: LEED buildings contribute to reduced greenhouse gas emissions and lower environmental impact.

Challenges:

Certification Costs: The certification process can be costly and time-consuming.

Maintenance Requirements: Maintaining LEED certification requires ongoing commitment to sustainability practices.

2. RE100 Initiative

Overview: The RE100 initiative is a global coalition of influential businesses committed to using 100% renewable electricity.

Benefits:

Corporate Leadership: Joining RE100 demonstrates corporate leadership in sustainability and climate action.

Collaborative Network: Members benefit from a network of like-minded companies and shared best practices.

Public Commitment: Publicly committing to 100% renewable energy enhances corporate reputation and stakeholder trust.

Challenges:

Implementation: Achieving 100% renewable energy can be challenging, particularly for companies with global operations.

Regulatory Variability: Different regulatory environments across countries can complicate the transition to renewable energy.

3. Science-Based Targets

Overview: Science-based targets provide companies with a clearly defined pathway to reduce greenhouse gas emissions in line with the Paris Agreement goals.

Benefits:

Climate Leadership: Setting science-based targets positions companies as leaders in climate action.

Investor Confidence: Clear and credible targets can attract investment and enhance investor confidence.

Operational Efficiency: Achieving these targets often leads to operational efficiencies and cost savings.

Challenges:

Target Setting: Developing and validating science-based targets requires rigorous data analysis and planning.

Implementation: Implementing strategies to meet these targets can be complex and resource-intensive.

Renewable Energy in Corporate Social Responsibility (CSR)

1. Corporate Social Responsibility (CSR) and Sustainability

Overview: CSR involves companies taking responsibility for their impact on society and the environment, with renewable energy playing a key role in these efforts.

Benefits:

Brand Reputation: Demonstrating a commitment to renewable energy enhances corporate reputation and brand loyalty.

Stakeholder Engagement: Engaging stakeholders, including customers, employees, and investors, around sustainability efforts fosters trust and collaboration.

Regulatory Compliance: Proactive sustainability initiatives help companies stay ahead of regulatory requirements and mitigate risks.

2. Environmental Impact Reduction

Overview: Renewable energy adoption helps companies reduce their environmental impact and carbon footprint.

Examples:

Microsoft: Microsoft has committed to being carbon negative by 2030 and is investing in renewable energy projects to achieve this goal.

Unilever: Unilever sources 100% of its electricity for its global operations from renewable sources, significantly reducing its greenhouse gas emissions.

Tesla: Tesla not only produces electric vehicles but also invests in renewable energy solutions, such as solar panels and battery storage, to power its operations sustainably.

3. Community Engagement and Development

Overview: Corporations can use renewable energy projects to engage with and support local communities.

Examples:

IKEA: IKEA has partnered with local communities to develop renewable energy projects, such as wind farms, which provide clean energy and economic benefits.

Starbucks: Starbucks has invested in community-based solar projects, providing clean energy to its stores and surrounding communities.

Siemens: Siemens has supported renewable energy education and training programs in communities where it operates, fostering local job creation and skills development.

Conclusion

Corporations and industries are increasingly recognizing the importance of renewable energy in their sustainability and CSR strategies. Through commitments to renewable energy, green certifications, and community engagement, companies are not only reducing their environmental impact but also enhancing their brand reputation and stakeholder relationships. While challenges remain, the growing corporate embrace of renewable energy is a positive sign of progress toward a more sustainable and resilient global economy. By continuing to invest in renewable energy and sustainability initiatives, corporations can lead the way in combating climate change and promoting a cleaner, greener future for all.

Chapter 17
Community and Grassroots Movements

Introduction to Community and Grassroots Movements

Community and grassroots movements have become powerful forces in promoting renewable energy and combating climate change. These movements are driven by local initiatives and community-owned renewable projects that empower citizens to take action and drive change from the ground up. This chapter explores the significant impact of these movements, the various models of community renewable energy projects, and the challenges and successes they encounter.

The Role of Community and Grassroots Movements

1. Empowering Local Communities

Overview: Grassroots movements empower local communities to take control of their energy sources and promote sustainable practices.

Benefits:

Local Ownership: Community-owned renewable energy projects provide direct benefits to local residents, including economic gains and energy security.

Engagement and Education: These movements raise awareness about renewable energy and climate change, fostering a culture of sustainability.

Resilience: Local projects enhance community resilience by reducing dependence on centralized fossil fuel-based energy systems.

2. Driving Policy Change

Overview: Grassroots movements often influence policy by advocating for supportive regulations and incentives for renewable energy.

Examples:

Lobbying and Advocacy: Grassroots organizations lobby local and national governments for policies that promote renewable energy and remove barriers to adoption.

Public Campaigns: Public awareness campaigns highlight the benefits of renewable energy, putting pressure on policymakers to act.

Legal Action: Some movements resort to legal action to challenge policies and practices that hinder renewable energy development.

3. Innovative Financing and Ownership Models

Overview: Community and grassroots movements have developed innovative financing and ownership models to fund renewable energy projects.

Models:

Community Shares: Residents can buy shares in local renewable energy projects, providing the necessary capital for development and ensuring community ownership.

Crowdfunding: Crowdfunding platforms allow individuals to contribute small amounts of money to support renewable energy projects.

Cooperatives: Energy cooperatives are member-owned organizations that develop, own, and manage renewable energy projects, distributing profits among members.

Examples of Community Renewable Energy Projects

1. Wind Power Projects

Overview: Community-owned wind power projects harness the power of local winds to generate electricity.

Examples:

Denmark: Denmark is a global leader in community wind projects, with many wind turbines owned by local cooperatives and residents.

Scotland: The Isle of Gigha in Scotland has a community-owned wind farm that generates significant income for the local community.

United States: In the U.S., community wind projects have been developed in states like Minnesota and Iowa, providing local benefits and supporting rural economies.

2. Solar Energy Projects

Overview: Solar energy projects are popular among communities due to the accessibility and scalability of solar technology.

Examples:

Germany: Germany's Energiewende has seen the rise of numerous community solar projects, where local residents invest in and benefit from solar installations.

Australia: The Repower Shoalhaven project in Australia is a community-owned solar farm that supplies clean energy to the local grid.

India: In India, community solar initiatives have brought electricity to remote villages, improving quality of life and economic opportunities.

3. Biomass and Bioenergy Projects

Overview: Community biomass projects utilize local organic materials to produce energy, supporting sustainable waste management.

Examples:

United Kingdom: The community of Ashton Hayes in the UK has a biomass heating system that provides renewable heat to local homes and businesses.

Kenya: In Kenya, community biogas projects convert agricultural and animal waste into clean energy for cooking and lighting.

Canada: The Cowichan Bio-Diesel Cooperative in Canada produces biodiesel from waste cooking oil, reducing waste and providing an alternative fuel source.

Challenges Faced by Community Renewable Energy Projects

1. Financial Barriers

Overview: Securing funding for community renewable energy projects can be challenging, particularly in low-income areas.

Challenges:

Upfront Costs: High initial capital costs for renewable energy installations can be a significant barrier for community projects.

Access to Financing: Limited access to financing options, such as loans and grants, can hinder project development.

Financial Risk: Community members may be reluctant to invest due to perceived financial risks and uncertainties.

2. Regulatory and Policy Hurdles

Overview: Regulatory and policy environments can either support or impede community renewable energy initiatives.

Challenges:

Permitting and Licensing: Complex and time-consuming permitting processes can delay or block community projects.

Grid Access: Obtaining access to the grid for small-scale renewable energy projects can be difficult, particularly in regions with monopolistic energy markets.

Policy Uncertainty: Inconsistent or unstable policy environments can create uncertainty for community investments in renewable energy.

3. Technical and Logistical Issues

Overview: Technical and logistical challenges can arise in the development and operation of community renewable energy projects.

Challenges:

Technical Expertise: Communities may lack the technical expertise needed to design, install, and maintain renewable energy systems.

Supply Chain: Securing reliable supply chains for equipment and materials can be challenging, particularly in remote areas.

Maintenance and Operation: Ensuring the long-term maintenance and efficient operation of renewable energy systems requires ongoing commitment and resources.

Success Stories and Best Practices

1. Best Practices for Community Engagement

Overview: Successful community renewable energy projects often involve strong community engagement and participation.

Strategies:

Inclusive Decision-Making: Involving all community members in the decision-making process fosters ownership and commitment to the project.

Education and Awareness: Providing education and raising awareness about the benefits and feasibility of renewable energy helps build community support.

Transparency and Communication: Maintaining transparent communication about project goals, progress, and challenges builds trust and confidence among community members.

2. Leveraging Partnerships

Overview: Partnerships with government agencies, NGOs, and private sector entities can enhance the success of community renewable energy projects.

Examples:

Public-Private Partnerships: Collaborating with private companies can provide technical expertise, financial resources, and market access.

NGO Support: Non-governmental organizations can offer funding, advocacy, and capacity-building support for community projects.

Government Programs: Government grants, subsidies, and technical assistance programs can reduce financial barriers and provide regulatory support.

3. Adaptive and Resilient Planning

Overview: Successful projects are adaptable and resilient, capable of responding to changing circumstances and challenges.

Strategies:

Flexible Design: Designing renewable energy systems that can be easily expanded or modified as needs and technologies change.

Risk Management: Implementing risk management strategies to address financial, technical, and operational risks.

Continuous Improvement: Regularly evaluating project performance and incorporating lessons learned to improve future initiatives.

Conclusion

Community and grassroots movements are essential drivers of the global transition to renewable energy. By empowering local communities, advocating for policy change, and developing innovative financing models, these movements are making significant contributions to a sustainable future. Despite facing financial, regulatory, and technical challenges, many community renewable energy projects have achieved remarkable success, providing valuable lessons and best practices for others to follow. The collective efforts of communities around the world demonstrate the power of grassroots action in addressing climate change and promoting renewable energy.

Chapter 18
The Future of Renewable Energy

Introduction to the Future of Renewable Energy

As the world grapples with the pressing need to combat climate change and secure sustainable energy sources, the future of renewable energy holds immense promise. This chapter offers a forward-looking perspective on potential breakthroughs, emerging technologies, and the long-term outlook for global energy systems. It delves into the innovations that could transform the energy landscape, the trends shaping the industry, and the vision for a renewable energy-powered future.

Potential Breakthroughs in Renewable Energy

1. Advanced Solar Technologies

Perovskite Solar Cells: Perovskite materials have shown remarkable efficiency improvements and could potentially surpass silicon-based solar cells. These cells are cheaper to produce and can be used in flexible applications, opening new possibilities for solar integration.

Tandem Solar Cells: Combining multiple types of solar cells (such as perovskite and silicon) in tandem can significantly boost overall efficiency, enabling higher energy yields from the same surface area.

Solar Windows and Building-Integrated Photovoltaics (BIPV): Innovations in transparent solar cells and BIPV allow for the integration of solar energy generation directly into building materials, such as windows and facades, turning structures into power generators.

2. Next-Generation Wind Energy

Floating Wind Turbines: Offshore wind farms are moving further into deeper waters with floating wind turbines. These turbines can access stronger and more consistent wind resources, significantly increasing potential energy output.

Vertical Axis Wind Turbines (VAWTs): Unlike traditional horizontal axis wind turbines, VAWTs can capture wind from any direction and are more suited to urban and small-scale applications, enhancing versatility and deployment options.

Airborne Wind Energy Systems: Using kites or drones to harness wind at higher altitudes, where it is stronger and more consistent, could revolutionize wind energy by providing continuous power generation.

3. Breakthroughs in Energy Storage

Solid-State Batteries: Promising higher energy density, faster charging times, and improved safety, solid-state batteries could replace conventional lithium-ion batteries in the near future.

Flow Batteries: Utilizing liquid electrolytes, flow batteries offer scalable and long-duration energy storage solutions, ideal for balancing grid supply and demand.

Hydrogen Storage: Hydrogen produced from renewable energy through electrolysis can be stored and used as a clean fuel for various applications, including transportation and power generation.

4. Innovations in Grid Integration and Management

Smart Grids: The development of smart grids, equipped with advanced sensors, communication technologies, and automated control systems, enhances the efficiency, reliability, and integration of renewable energy sources.

Microgrids: Localized energy grids that can operate independently or in conjunction with the main grid, microgrids provide energy security, resilience, and the ability to integrate diverse renewable energy sources.

Blockchain for Energy Trading: Blockchain technology can enable peer-to-peer energy trading, allowing consumers to buy and sell excess renewable energy directly, fostering decentralized and efficient energy markets.

Emerging Technologies Shaping the Future

1. Green Hydrogen

Production: Green hydrogen, produced using renewable energy to split water into hydrogen and oxygen, offers a zero-emission fuel alternative for various sectors.

Applications: Hydrogen can be used in fuel cells for transportation, as a feedstock for industrial processes, and in power generation, providing a versatile and clean energy carrier.

Challenges: Scaling up green hydrogen production and developing infrastructure for storage, transport, and utilization are critical challenges to address.

2. Marine Energy

Wave Energy: Harnessing the power of ocean waves through various technologies, such as point absorbers and oscillating water columns, wave energy has the potential to provide consistent and reliable power.

Tidal Energy: Utilizing the predictable movements of tides, tidal energy systems, including tidal barrages and underwater turbines, offer another reliable renewable energy source.

Ocean Thermal Energy Conversion (OTEC): Exploiting the temperature differences between warm surface water and cold deep water, OTEC systems can generate continuous power in tropical regions.

3. Bioenergy and Waste-to-Energy

Advanced Biofuels: Innovations in biofuel production, including algae-based biofuels and cellulosic ethanol, provide sustainable alternatives to fossil fuels for transportation.

Anaerobic Digestion: Converting organic waste into biogas through anaerobic digestion offers a renewable energy source while addressing waste management issues.

Waste-to-Energy Plants: Technologies that convert municipal solid waste into electricity and heat, such as incineration and gasification, contribute to renewable energy generation and waste reduction.

Long-Term Outlook for Global Energy Systems

1. Decarbonization and Energy Transition

Net-Zero Emissions: Achieving net-zero emissions by mid-century is a key goal for many countries, requiring a massive scale-up of renewable energy deployment and the phase-out of fossil fuels.

Electrification: Electrifying sectors such as transportation, heating, and industry using renewable energy sources is crucial for reducing greenhouse gas emissions.

Energy Efficiency: Improving energy efficiency across all sectors reduces overall energy demand, complementing the shift to renewable energy.

2. Global Collaboration and Policy Support

International Agreements: Agreements such as the Paris Agreement provide a framework for global collaboration on climate action and renewable energy deployment.

Government Policies: Supportive policies, including subsidies, tax incentives, and renewable energy mandates, are essential for accelerating the adoption of renewable energy technologies.

Private Sector Investment: Continued investment from the private sector, driven by corporate sustainability goals and the economic benefits of renewable energy, will play a critical role in the transition.

3. Socio-Economic Impacts

Job Creation: The renewable energy sector is expected to create millions of jobs worldwide, contributing to economic growth and social development.

Energy Access: Expanding renewable energy infrastructure, particularly in developing countries, can provide reliable and affordable energy access, improving quality of life and supporting economic development.

Community Empowerment: Community-owned renewable energy projects and grassroots movements empower local communities, fostering sustainable development and resilience.

Conclusion

The future of renewable energy is bright, with numerous potential breakthroughs, emerging technologies, and transformative trends on the horizon. As the world moves towards a sustainable energy future, the continued innovation and deployment of renewable energy technologies will be essential. Achieving global climate goals, ensuring energy security, and promoting socio-economic development all hinge on the successful integration and expansion of renewable energy systems. By harnessing the power of the sun, wind, water, and organic matter, humanity can build a resilient and sustainable energy future for generations to come.

Chapter 19
Renewable Energy and Developing Nations

Introduction to Renewable Energy in Developing Nations

Developing nations face unique challenges and opportunities in their pursuit of sustainable development. Renewable energy presents a transformative potential for these countries, offering not only access to electricity but also a pathway to economic development and improved quality of life. This chapter explores the critical role of renewable energy in developing nations, examining the obstacles they face, the innovative solutions being implemented, and the broader implications for global energy equity.

The Importance of Renewable Energy in Developing Nations

1. Access to Electricity

Energy Poverty: Many developing nations suffer from significant energy poverty, where large segments of the population lack reliable access to electricity. Renewable energy can bridge this gap, providing decentralized and scalable solutions.

Off-Grid Solutions: Solar home systems, mini-grids, and other off-grid renewable energy solutions can bring electricity to remote and rural areas where traditional grid extension is impractical or too expensive.

Improved Quality of Life: Access to electricity improves quality of life by enabling lighting, refrigeration, communication, and the use of various household appliances. It also supports education and healthcare services by powering schools and clinics.

2. Economic Development

Job Creation: The renewable energy sector offers significant employment opportunities, from manufacturing and installation to operation and maintenance of energy systems. This can stimulate local economies and reduce poverty.

Entrepreneurship: Renewable energy can spur entrepreneurship by providing power for small businesses and agricultural activities, fostering economic diversification and growth.

Industrialization: Reliable and sustainable energy is crucial for industrial development, enabling the establishment and expansion of industries that require stable power supplies.

3. Environmental and Health Benefits

Reduced Pollution: Transitioning from traditional biomass and fossil fuels to clean energy sources reduces air pollution, which is a major health hazard in many developing countries.

Climate Resilience: Renewable energy systems, particularly decentralized ones, enhance climate resilience by reducing dependence on imported fuels and mitigating the impact of energy supply disruptions.

Sustainable Resource Management: Utilizing local renewable resources, such as solar, wind, and biomass, promotes sustainable resource management and reduces environmental degradation.

Challenges to Renewable Energy Adoption

1. Financial Barriers

High Initial Costs: The upfront capital costs for renewable energy projects can be prohibitive for many developing nations, where financial resources are often limited.

Access to Finance: Limited access to affordable financing options, such as loans and grants, hampers the ability of individuals and communities to invest in renewable energy solutions.

Economic Instability: Economic instability and high levels of debt in some developing nations can deter investment in renewable energy projects.

2. Technical and Infrastructure Challenges

Lack of Technical Expertise: A shortage of skilled professionals in the renewable energy sector can impede the planning, installation, and maintenance of energy systems.

Inadequate Infrastructure: Poor infrastructure, including weak or nonexistent power grids, transportation networks, and communication systems, can obstruct the deployment of renewable energy technologies.

Resource Assessment: Accurate assessment and mapping of renewable energy resources (such as solar irradiance, wind patterns, and geothermal potential) are often lacking, making it difficult to plan effective projects.

3. Policy and Regulatory Hurdles

Policy Uncertainty: Inconsistent or unclear policies and regulations can create uncertainty for investors and project developers.

Subsidies for Fossil Fuels: Continued subsidies for fossil fuels in many developing nations make renewable energy less competitive and attractive.

Bureaucratic Obstacles: Complex and time-consuming administrative procedures can delay or prevent the implementation of renewable energy projects.

Opportunities and Innovative Solutions

1. International Collaboration and Funding

Development Aid: International development aid and funding from organizations such as the World Bank, the International Renewable Energy Agency (IRENA), and various NGOs support renewable energy projects in developing nations.

Public-Private Partnerships: Collaborations between governments, private companies, and international organizations can mobilize resources and expertise for large-scale renewable energy initiatives.

Climate Finance: Mechanisms such as the Green Climate Fund (GCF) provide financial support for climate mitigation and adaptation projects, including renewable energy development.

2. Innovative Business Models

Pay-As-You-Go (PAYG) Solar: PAYG models allow customers to pay for solar energy systems in affordable installments, making renewable energy accessible to low-income households.

Microfinance: Microfinance institutions provide small loans to individuals and communities to invest in renewable energy solutions, fostering local ownership and empowerment.

Community Ownership: Community-owned renewable energy projects, such as cooperatives, ensure that the benefits of clean energy are shared among local residents, promoting social and economic equity.

3. Technological Adaptations

Modular and Scalable Solutions: Modular renewable energy systems, such as solar home kits and mini-grids, can be scaled up as demand grows, providing flexible and adaptable energy solutions.

Hybrid Systems: Combining different renewable energy sources (e.g., solar and wind) with battery storage enhances reliability and ensures a continuous power supply.

Smart Technologies: The integration of smart technologies, such as remote monitoring and control systems, improves the efficiency and reliability of renewable energy systems in remote and underserved areas.

Case Studies of Renewable Energy in Developing Nations

1. Solar Energy in Sub-Saharan Africa

M-KOPA Solar (Kenya): M-KOPA Solar provides affordable solar home systems to off-grid households in Kenya using a PAYG model. Customers make small daily payments via mobile money, enabling widespread access to clean energy.

Lighting Africa (Multiple Countries): The World Bank's Lighting Africa program has improved access to affordable, reliable, and modern off-grid lighting products, benefiting millions of people across Sub-Saharan Africa.

2. Wind Energy in South Asia

Wind Power in India: India has made significant strides in wind energy development, particularly in states like Tamil Nadu and Gujarat. Government policies, financial incentives, and international investments have driven this growth.

Community Wind Projects (Sri Lanka): Community-based wind energy projects in Sri Lanka provide electricity to rural areas, enhancing energy security and supporting local development.

3. Hydropower in Latin America

Micro-Hydro Projects (Nepal): Nepal has successfully implemented numerous micro-hydro projects in remote mountainous regions, providing reliable electricity and supporting local livelihoods.

Run-of-the-River Hydropower (Costa Rica): Costa Rica's focus on small-scale, environmentally friendly run-of-the-river hydropower projects has contributed to its status as a global leader in renewable energy.

4. Biomass and Bioenergy in Southeast Asia

Biogas Systems (Vietnam): Vietnam's National Biogas Program has installed thousands of biogas digesters, converting agricultural waste into clean energy and reducing reliance on traditional biomass.

Rice Husk Power (Cambodia): In Cambodia, rice husk gasification plants generate electricity from agricultural residues, providing a renewable energy source and reducing waste.

Conclusion

Renewable energy holds the key to addressing some of the most pressing challenges faced by developing nations, including energy poverty, economic development, and environmental sustainability. Despite significant barriers, innovative solutions and international collaboration are driving progress and demonstrating the transformative potential of clean energy. By harnessing their abundant renewable resources and leveraging supportive policies and financing mechanisms, developing nations can pave the way towards a sustainable and equitable energy future. The experiences and successes of these countries provide valuable lessons and inspiration for the global transition to renewable energy.

Chapter 20
Conclusion and Call to Action

Summary of Key Points

Throughout this book, we have explored the critical role of renewable energy in combating climate change and driving sustainable development. Each chapter has delved into various aspects of renewable energy, from its fundamental concepts to its technological advancements, economic implications, and societal benefits. Here, we summarize the key points discussed and provide a call to action for individuals, businesses, and governments to actively participate in the renewable energy revolution.

1. Understanding Renewable Energy

We began with an introduction to the different types of renewable energy sources—solar, wind, hydro, geothermal, and biomass. These sources offer clean, sustainable alternatives to fossil fuels and are pivotal in reducing greenhouse gas emissions.

2. Historical Perspective

A historical overview highlighted the evolution of energy consumption, from the Industrial Revolution's reliance on coal to the gradual shift towards renewable energy driven by technological advancements and growing environmental awareness.

3. The Science of Climate Change

An in-depth examination of climate change science underscored the urgency of transitioning to renewable energy. The greenhouse effect, carbon emissions, and human activities' role in climate change were explained, emphasizing why renewable energy is critical for mitigation efforts.

4. Solar Power

We explored solar energy technologies, including photovoltaic cells and solar thermal systems, and their potential to become major energy sources. Innovations in solar technology were highlighted, showcasing the industry's rapid advancements.

5. Wind Energy

Detailed coverage of wind energy included the mechanics of wind turbines, the development of onshore and offshore wind farms, and the future potential of wind power. The benefits and challenges of wind energy were discussed, emphasizing its importance in the renewable energy mix.

6. Hydro Power

The chapter on hydropower delved into large-scale dams and micro-hydro projects, discussing their environmental and social impacts and their role in providing a stable and reliable energy source.

7. Geothermal Energy

Geothermal energy's potential was examined, highlighting its ability to provide a consistent energy supply. The processes of harnessing geothermal energy and its applications in power generation and heating were explored.

8. Biomass Energy

Biomass energy was discussed, including its sources, conversion processes, and sustainability. The role of biomass in the renewable energy landscape and its potential to reduce waste and generate energy were emphasized.

9. Energy Storage and Grid Integration

The challenges and solutions for integrating renewable energy into existing grids were explored, with a focus on energy storage technologies and smart grids. The importance of reliable and efficient grid integration for maximizing renewable energy utilization was highlighted.

10. Policy and Legislation

Global policies and legislation promoting renewable energy were reviewed, including international agreements, national policies, and incentives. The role of government support in accelerating renewable energy adoption was emphasized.

11. Economic Impacts

The economic aspects of renewable energy were explored, including job creation, investment opportunities, and the cost-benefit analysis of transitioning from fossil fuels. The chapter highlighted renewable energy's potential to drive economic growth.

12. Environmental and Social Benefits

The environmental and social benefits of renewable energy, such as reduced air pollution, improved public health, and enhanced energy security, were discussed. The chapter underscored renewable energy's positive impact on communities and the environment.

13. Case Studies of Successful Projects

Real-world examples of successful renewable energy projects demonstrated the practical applications and benefits of renewable energy in various contexts. These case studies provided inspiration and lessons for future initiatives.

14. Technological Innovations

Advancements in renewable energy technologies were examined, showcasing the latest innovations in solar panels, wind turbine designs, and energy storage solutions. The chapter highlighted the ongoing progress in the renewable energy sector.

15. Barriers to Adoption

The barriers to the widespread adoption of renewable energy were discussed, including technological limitations, economic challenges, and political resistance. Strategies for overcoming these obstacles were explored.

16. The Role of Corporations and Industry

The chapter analyzed how corporations and industries are embracing renewable energy through sustainability initiatives, green certifications, and corporate social responsibility. The private sector's role in driving renewable energy adoption was emphasized.

17. Community and Grassroots Movements

The role of community and grassroots movements in promoting renewable energy was examined, showcasing how local initiatives and community-owned projects are driving change from the ground up.

18. The Future of Renewable Energy

A forward-looking perspective on the future of renewable energy explored potential breakthroughs, emerging technologies, and the long-term outlook for global energy systems. The chapter highlighted the transformative potential of renewable energy innovations.

19. Renewable Energy in Developing Nations

The unique challenges and opportunities for renewable energy in developing nations were explored, emphasizing its role in providing access to electricity, driving economic development, and improving quality of life.

Call to Action

The transition to renewable energy is not just a technological or economic challenge; it is a societal imperative. The stakes are high, and the need for action is urgent. Here, we outline specific actions that individuals, businesses, and governments can take to contribute to the renewable energy revolution.

1. Individuals

Reduce Energy Consumption: Adopt energy-efficient practices at home and in daily life. Simple actions like using energy-efficient appliances, insulating homes, and reducing unnecessary energy use can make a significant impact.

Support Renewable Energy: Choose green energy options if available from your utility provider. Invest in renewable energy technologies such as solar panels for your home.

Advocate for Change: Support policies and initiatives that promote renewable energy. Engage in community efforts and advocate for sustainable energy practices within your local area.

2. Businesses

Invest in Renewable Energy: Incorporate renewable energy sources into your operations. This can include installing solar panels, purchasing renewable energy credits, or investing in renewable energy projects.

Sustainability Initiatives: Implement comprehensive sustainability initiatives that include energy efficiency measures, waste reduction, and renewable energy integration.

Corporate Social Responsibility: Adopt corporate social responsibility practices that prioritize environmental sustainability and contribute to the broader goal of combating climate change.

3. Governments

Policy and Regulation: Enact and enforce policies that support renewable energy development. This includes setting renewable energy targets, providing incentives, and phasing out fossil fuel subsidies.

Infrastructure Investment: Invest in the necessary infrastructure for renewable energy, such as smart grids, energy storage systems, and public transportation powered by clean energy.

International Collaboration: Participate in international agreements and collaborations to share knowledge, technology, and resources for global renewable energy development.

Conclusion

The renewable energy revolution is not just a pathway to mitigating climate change; it is a gateway to a sustainable, equitable, and prosperous future. By embracing renewable energy, we can address environmental challenges, foster economic growth, and improve the quality of life for people worldwide. The transition to renewable energy requires concerted efforts from all sectors of society—individuals, businesses, and governments alike.

The journey towards a renewable energy future is filled with challenges, but it is also brimming with opportunities. Together, we can harness the power of the sun, wind, water, and organic matter to build a resilient and sustainable world.

The time to act is now. Let us commit to making renewable energy the cornerstone of our energy systems and work collectively to secure a brighter future for generations to come.

Closing of the book

As we reach the end of "The Role of Renewable Energy in Combating Climate Change," it is clear that the path to a sustainable future lies in our collective ability to embrace and advance renewable energy technologies. This journey is not just about technological innovation; it is about rethinking our relationship with the environment and each other. It is about ensuring a future where clean, sustainable energy is accessible to all, where economic opportunities abound, and where the devastating impacts of climate change are mitigated.

The Urgency of Now

The scientific consensus on climate change is unequivocal. The window of opportunity to avert the worst impacts of global warming is rapidly closing. The actions we take today will determine the health and prosperity of future generations. Renewable energy stands at the forefront of these actions, offering a powerful tool to decarbonize our energy systems and create a resilient, low-carbon economy.

The Promise of Renewable Energy

Renewable energy is more than a technological solution; it is a catalyst for broader societal change. It promises a future where energy production does not come at the expense of environmental degradation, where economic growth is decoupled from carbon emissions, and where energy security and equity are prioritized. By harnessing the power of the sun, wind, water, and organic matter, we can build a world that is not only sustainable but also more just and equitable.

A Collaborative Effort

The transition to renewable energy requires the concerted efforts of individuals, businesses, and governments. Each of us has a role to play:

Individuals: We can make informed choices about our energy use, advocate for renewable energy policies, and support community renewable energy projects.

Businesses: Companies can invest in renewable energy technologies, adopt sustainable practices, and lead by example through corporate social responsibility initiatives.

Governments: Policymakers can create supportive regulatory frameworks, provide incentives for renewable energy development, and invest in the necessary infrastructure to facilitate the transition.

Moving Forward

As we move forward, it is essential to keep the momentum going. The renewable energy revolution is well underway, but there is still much work to be done. Continued research and development, investment in new technologies, and the implementation of forward-thinking policies are all crucial to ensuring the success of this transition.

Moreover, it is vital to address the barriers that still exist. Technological, economic, and political challenges must be overcome to fully realize the potential of renewable energy. By fostering collaboration and innovation, we can find solutions to these obstacles and accelerate the adoption of clean energy.

A Vision for the Future

Imagine a future where cities are powered by renewable energy, where transportation systems are clean and efficient, and where communities thrive on sustainable practices. This is not a distant dream but a tangible reality that we can achieve through collective action and unwavering commitment.

Final Thoughts

The role of renewable energy in combating climate change cannot be overstated. It is a critical component of the broader strategy to create a sustainable and resilient world. By understanding the importance of renewable energy, advocating for its adoption, and actively participating in the transition, we can make a profound impact on the health of our planet and the well-being of its inhabitants.

In closing, let us remember that the journey towards a sustainable future is a shared one. Each step we take, no matter how small, contributes to the larger goal of a cleaner, greener, and more equitable world. Together, we have the power to drive the renewable energy revolution and ensure a brighter future for generations to come.

Thank you for joining us on this journey. Let us continue to work together, innovate, and inspire others to embrace the promise of renewable energy. The future is in our hands.

Sincerely,

FAISAL JAMIL

Thank You for Reading

Thank you for choosing "The Role of Renewable Energy in Combating Climate Change." We hope this book has provided you with valuable insights and inspiration for a sustainable future. Your feedback is incredibly important to us and helps guide future readers.

If you found this book informative and engaging, please consider leaving a positive review on Amazon. Your support is greatly appreciated and contributes to spreading awareness about the critical role of renewable energy in combating climate change. Thank you for being part of this journey towards a greener, cleaner world!

www.ingramcontent.com/pod-product-compliance
Lightning Source LLC
Chambersburg PA
CBHW071921210526
45479CB00002B/510